Experiments for Electronic Principles

Second Edition

(A laboratory manual for use with *Electronic Principles,* second edition)

Albert Paul Malvino, Ph.D.

Foothill College
Los Altos Hills, California
West Valley College
Saratoga, California

Gregory F. Johnson

INELEC, Electricité et Electronique
Boumerdes, Algeria

Gregg Division

McGraw-Hill Book Company

New York St. Louis Dallas San Francisco Auckland
Bogotá Düsseldorf Johannesburg London Madrid
Mexico Montreal New Delhi Panama Paris
São Paulo Singapore Sydney Tokyo Toronto

Let your imagination go, guiding it by principle and experiment. Nature is your best friend and critic if you only let her intimations fall unbiased on your mind; nothing is so good as an experiment that sets an error right and improves your understanding.

Faraday

Experiments for Electronic Principles, Second Edition

ISBN 0-07-039868-2

5 6 7 8 9 0 D O D O   8 5 4 3 2 1

The editors for this book were George Horesta and Myrna Breskin, the designer was Eileen Thaxton, the cover designer was Blaise Zito Assoc., and its production was supervised by S. Steven Canaris. It was set in Baskerville by Kingsport Press. Printed and bound by R. R. Donnelley & Sons Company.

contents

preface

You can know all the theory in the world, but it's useless if it won't work in real life. This laboratory manual contains 54 experiments to help you learn the theory of *Electronic Principles* well enough to make it work in the laboratory.

This second edition reflects changes in the textbook. Now included are experiments on the three types of rectifiers, capacitor-input filters, voltage multipliers, zener regulators, slew rate and power bandwidth, thyristors, three-terminal IC regulators, active diode circuits, active filters, and other op-amp circuits. For your convenience, the Appendix contains data sheets for all diodes, transistors, and linear ICs used in the experiments.

Each experiment is straightforward. *Required reading* lists the pages in the textbook that must be covered before attempting the experiment. *Equipment* lists the parts and instruments needed. *Procedure* gives each step in the experiment. Finally, the *data* section contains the tables to be filled in during the course of the experiment. The *Questions for Experiments* appear grouped after the last experiment. Each question section is a test to let teacher and student find out how much was learned.

You will find these experiments interesting and at times exciting. They prove, expand, and dramatize the theory of *Electronic Principles;* they make it come alive. If you will try to learn all you can while doing these experiments, you will know a great deal of practical electronics when finished.

Albert Paul Malvino

Gregory F. Johnson

1 current and voltage sources

THEORY

Dc voltage sources deliver a constant voltage when load resistance is fixed. In the ideal approximation, the voltage stays fixed no matter how much current flows. The second approximation is more realistic; it includes an internal resistance R. Only when large currents flow do we need to worry about the voltage drop across this internal resistance. When the load resistance is at least 100 times greater than the internal resistance of the source, we get less than 1 percent error by neglecting the internal resistance.

Dc current sources deliver a constant current when load resistance is fixed. Ideally, the current stays fixed no matter what the load resistance. To a second approximation, however, the current source has a high internal resistance. Only when the load resistance is high, do we have to take the internal resistance into account. When the load resistance is at least 100 times smaller than the source resistance, we can neglect the internal resistance with less than 1 percent error.

REQUIRED READING

Pages 1–8

EQUIPMENT

1 power supply: 10 V
6 ½-W resistors: 10 Ω, 47 Ω, 100 Ω, 470 Ω, 1 kΩ, 10 kΩ
1 VOM (volt-ohm-milliammeter)

PROCEDURE

1. The circuit left of the AB terminals in Fig. 1*a* represents a voltage source and its internal resistance R. Connect the circuit of Fig 1*a* using the values of R given in Table 1. For each R value, measure and record V_L.
2. The circuit left of the AB terminals in Fig. 1*b* acts like a current source

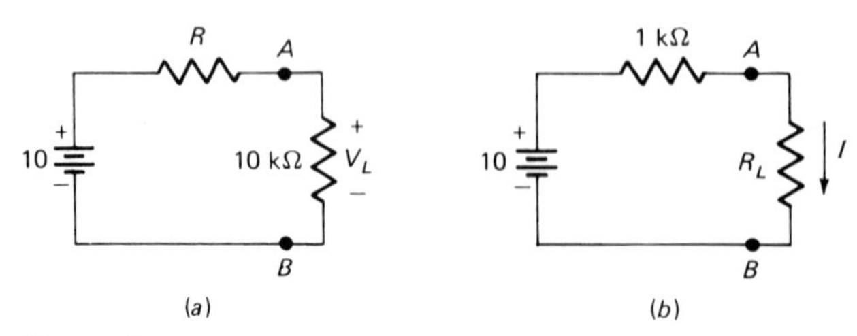

Figure 1

under certain conditions. Connect the circuit of Fig. 1*b* using the R_L values given in Table 2. For each R_L value, measure and record I.

DATA

Table 1. Voltage Source

R	V_L
0	10 V
10 Ω	10 V
100 Ω	9.91 V
470 Ω	9.55 V

Table 2. Current Source

R_L	I
0	10 mA
10 Ω	9.78 mA
47 Ω	9.51 mA
100 Ω	9.06 mA

2 *Thevenin's theorem*

REQUIRED READING

Pages 9–11

EQUIPMENT

1 power supply: 15 V
7 ½-W resistors: 470 Ω, two 1 kΩ, two 2.2 kΩ, two 4.7 kΩ
1 potentiometer: 5 kΩ

PROCEDURE

1. With R_L disconnected in Fig. 2*a*, work out the Thevenin voltage V_{TH} and the Thevenin resistance R_{TH}. Record these values in Table 1.
2. With the Thevenin values just found, calculate the load voltage V_L across an R_L of 1 kΩ (see Fig. 2*b*). Record V_L in Table 2.

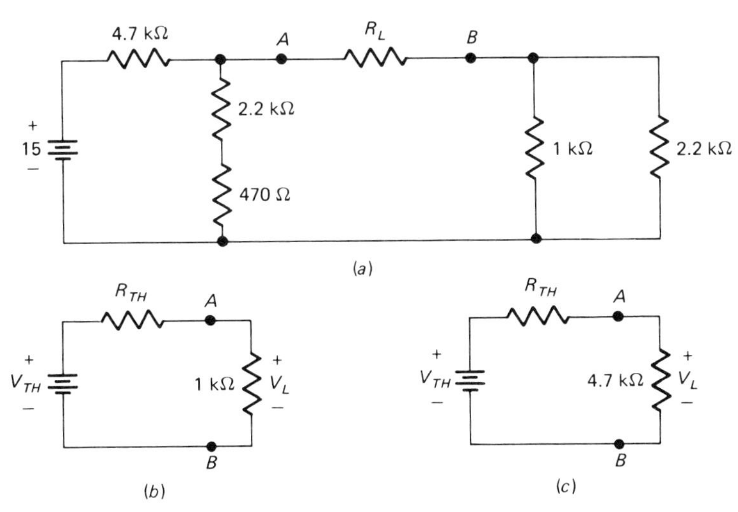

Figure 2

3. Also calculate the load voltage V_L for an R_L of 4.7 kΩ as shown in Fig. 2*c*. Record the calculated V_L in Table 2.
4. Connect the circuit of Fig. 2*a*, leaving out R_L.
5. Measure V_{TH} and record the value in Table 1.
6. Replace the 15-V source by a short circuit. Measure the resistance between the *AB* terminals using a convenient resistance range of the VOM. Record R_{TH} in Table 1. Now, replace the short by the 15-V source.
7. Connect a load resistance R_L of 1 kΩ between the *AB* terminals of Fig. 2*a*. Measure and record load voltage V_L (Table 2).
8. Change the load resistance from 1 kΩ to 4.7 kΩ. Measure and record the new load voltage.
9. Find R_{TH} by the matched-load method; that is, use the potentiometer as a variable resistance between the *AB* terminals of Fig. 2*a*. Vary resistance until load voltage drops to half of the measured V_{TH}. Then, disconnect the load resistance and measure its resistance with the VOM. This value should agree with R_{TH} found in Step 6. Yes

DATA

Table 1. Thevenin Values

Calculated V_{TH}	5.43 V
Calculated R_{TH}	2390 Ω
Measured V_{TH}	5.48
Measured R_{TH}	2.36 kΩ

Table 2. Load Voltages

Calculated V_L for 1 kΩ	1.60 V
Calculated V_L for 4.7 kΩ	3.60 V
Measured V_L for 1 kΩ	1.62
Measured V_L for 4.7 kΩ	3.62

3 *the diode curve*

REQUIRED READING

Pages 25–43

EQUIPMENT

1 power supply: adjustable from at least 1 to 15 V
1 diode: 1N914 (or almost any small-signal silicon diode)
3 ½-W resistors: 220 Ω, 1 kΩ, 100 kΩ

PROCEDURE

1. A fast diode test is the ohmmeter check. Using the VOM as an ohmmeter, measure the diode's forward resistance on the ×1 range. Next, measure the diode's reverse resistance on the highest ohmmeter range. If the diode is all right, you should have a low forward resistance and a high reverse resistance.
2. Connect the circuit of Fig. 3*a*. Adjust the voltage source until v_L equals 10 V. Measure diode voltage v and record this value in Table 1.

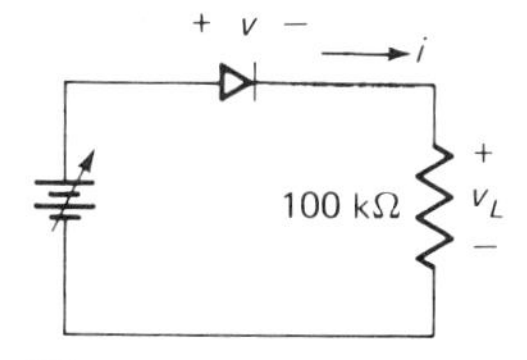

Figure 3a

3. Connect the circuit of Fig. 3*b*. Adjust the source until the load voltage equals 10 V. Measure diode voltage v and record in Table 1.
4. Reverse the polarity of the diode in Fig. 3*b*. With source adjusted to 10 V, measure load voltage v_L. Then measure diode voltage v. Record both values in Table 1.

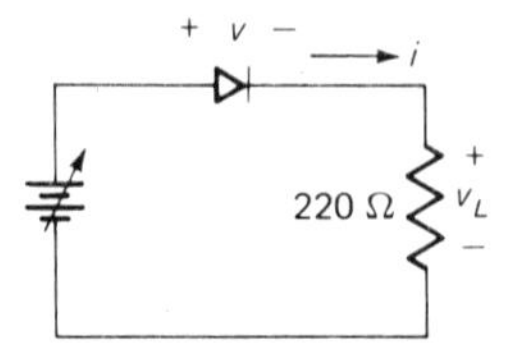

Figure 3b

5. Connect the circuit of Fig. 3*c*. Adjust the voltage source to produce each load voltage shown in Table 2. For each value, measure and record diode voltage v.

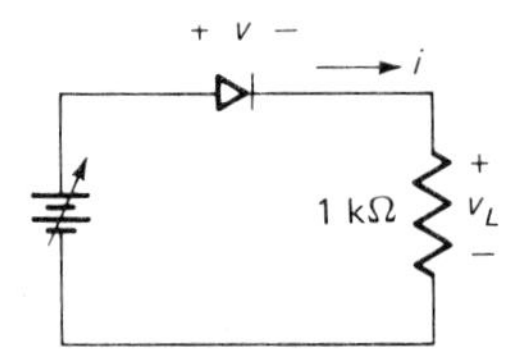

Figure 3c

6. The current through the diode has the same value as the current through the load resistance in Fig. 3*a*, *b*, and *c*. Use Ohm's law to calculate the value of i in Tables 1 and 2.
7. The dc resistance of a diode is defined as the ratio of direct voltage across the diode to the direct current through the diode. In symbols,

$$R = \frac{v}{i}$$

   Calculate and record each R value in Tables 1 and 2.
8. Graph the diode curve (i versus v) using the data of Tables 1 and 2.
9. If a curve tracer is available, use it to display the forward diode curve.

DATA

Table 1. Low and High Currents

R_L	V_L	v	i	R
100 kΩ	10	.508v	100μA	5kΩ
220 Ω	10	.804v	45.5mA	17.7Ω

R_L	V_L	V	(diode reversed)
220Ω	0v	10v	

Table 2. $R_L = 1\ \text{k}\Omega$

V_L	V	i	R
0	0v	0A	(no entry)
0.5	.57v	500μA	1.14kΩ
1	.60v	1mA	600Ω
2	.64v	2mA	320Ω
4	.67v	4mA	167.5Ω
6	.69v	6mA	115Ω
8	.70v	8mA	87.5Ω
10	.71v	10mA	71Ω

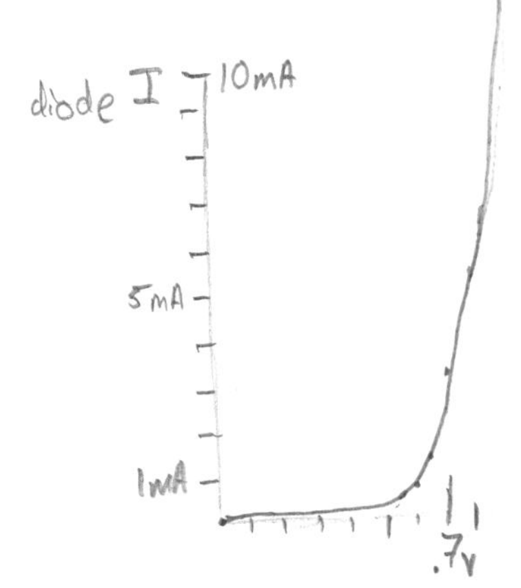

CURRENT VERSUS VOLTAGE of A DIODE

4 diode approximations

REQUIRED READING

Pages 43–48

EQUIPMENT

1 power supply: adjustable from at least 1 to 15 V
1 diode: 1N914 (or almost any small-signal silicon diode)
3 ½-W resistors: two 220 Ω, 470 Ω
1 VOM

PROCEDURE

1 Connect the circuit of Fig. 4*a*. Adjust the source to set up a current of 10 mA through the diode. (You can verify $i = 10$ mA by using the VOM as an ammeter or by measuring 2.2 V across the 220-Ω resistor.)

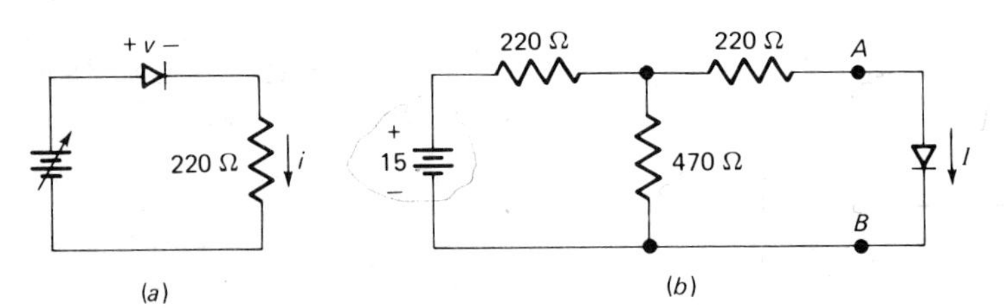

Figure 4

2. Measure diode voltage v and record this value in Table 1.
3. Adjust the source to get 50 mA. Measure and record v (Table 1).
4. In this experiment, we will let the knee voltage be the diode voltage corresponding to a diode current of 10 mA. Record the knee voltage in Table 2. (It should be in the vicinity of 0.7 V.)
5. Calculate the bulk resistance using

$$r_B = \frac{\Delta v}{\Delta i}$$

where Δv and Δi are the changes in voltage and current in Table 1. Record r_B in Table 2.

6. Connect the circuit of Fig. 4*b*. Get the direct current I through the diode by either of these methods: (1) using the VOM as an ammeter or (2) measuring the voltage across the 220-Ω resistance and calculating I.
   Record I under "Experimental I" in Table 3. (Because of limited choice of current ranges on some VOMs, method 2 often gives a more accurate value of I.)
7. Work out the value of I in Fig. 4*b* as follows: Thevenize the circuit left of the AB terminals. Then calculate I with the ideal, second, and third approximations of the diode (use the V_{knee} and r_B of Table 2).
8. Record the three calculated values of I in Table 3.

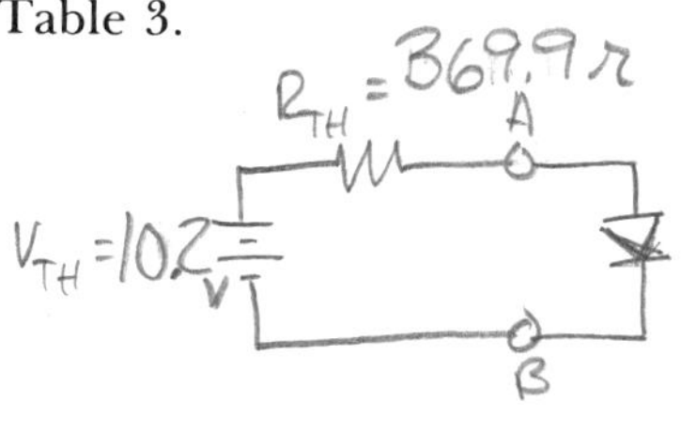

DATA

Table 1. Two Points on Forward Curve

i	*v*
10 mA	.727V
50 mA	.821V

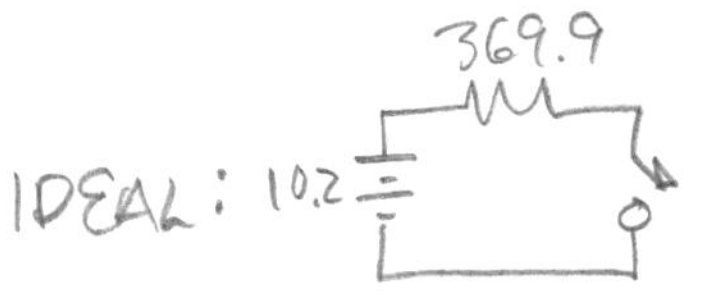

Table 2. Diode Values

V_{knee}	.727V
r_B	2.35Ω

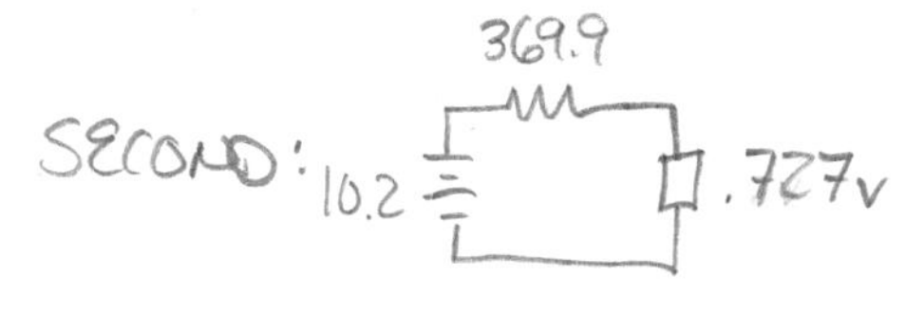

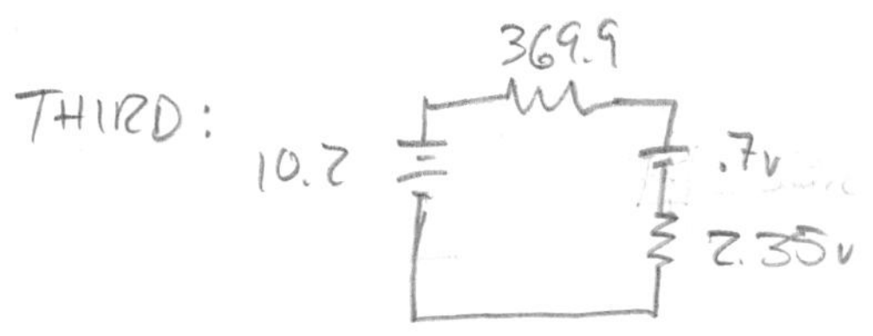

Table 3. Diode Current in Fig. 6*b*

Experimental I	26.0 mA
Ideal I	27.6 mA
Second I	25.6 mA
Third I	25.5 mA

5 *the zener diode*

REQUIRED READING

Pages 53–57

EQUIPMENT

1 power supply: adjustable from 1 to 15 V
1 zener diode: 1N753
1 ½-W resistor: 180 Ω
1 VOM (and a DVM if available)

PROCEDURE

1. Measure the diode's forward resistance on the ×1 range. Next, measure the reverse resistance on the ×1000 range.
2. Connect the circuit of Fig. 5.

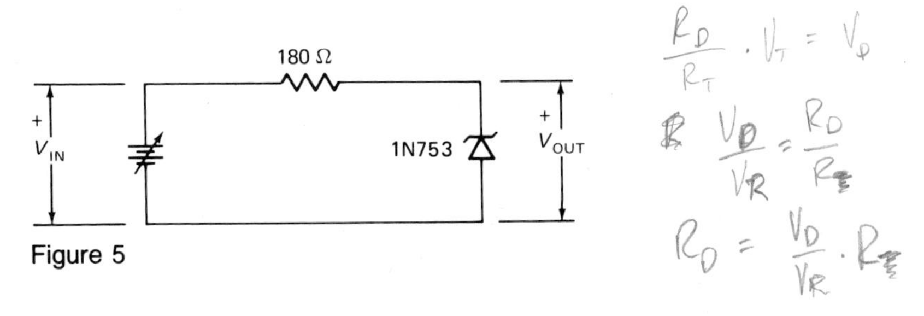

Figure 5

3. Measure and record the output voltage for each input voltage shown in Table 1. (Use a DVM if available.)
4. Calculate and record the zener current for each output voltage listed in Table 1.
5. The 1N753 has these specifications: $P_{Z(max)} = 400$ mW, $V_Z = 6.2$ V ± 10 percent, and $I_{ZT} = 20$ mA. What is the maximum allowable zener current?
6. With the data of Table 1, calculate the zener impedance in the vicinity of $V_{OUT} = 10$ V. (Use $Z_Z = \Delta V/\Delta I$ between a V_{OUT} of 8 to 12 V.)

.797 V

7. Reverse the polarity of the zener diode. Measure V_{OUT} for a V_{IN} of 15 V.
8. If a curve tracer is available, display the forward and reverse zener curves.

DATA

Table 1.

V_{IN}	V_{OUT}	I_Z
0	0	0
2	2	0
4	4	0
6	6	0
8	6.27	9.6 mA
10	6.29	20.6 mA
12	6.31	31.6 mA
15	6.34	48.0 mA

1N4735 ZENER
$V_Z = 6.2V$
$I_{ZT} = 41mA$
$P_D = 1W$

$P = I \cdot E$
$I = \frac{E}{P}$
$I = \frac{P}{E} = \frac{1W}{6.2V} = 161 mA$

6 *rectifier circuits*

REQUIRED READING

Pages 64–71

EQUIPMENT

1 center-tapped transformer, 12.6 V (Triad F-25X or equivalent) with fused line cord
4 silicon diodes: 1N4001 (or equivalent)
1 ½-W resistor: 1 kΩ
1 VOM (or DVM if available)
1 oscilloscope

PROCEDURE

1. Connect the half-wave rectifier shown in Fig. 6*a*.

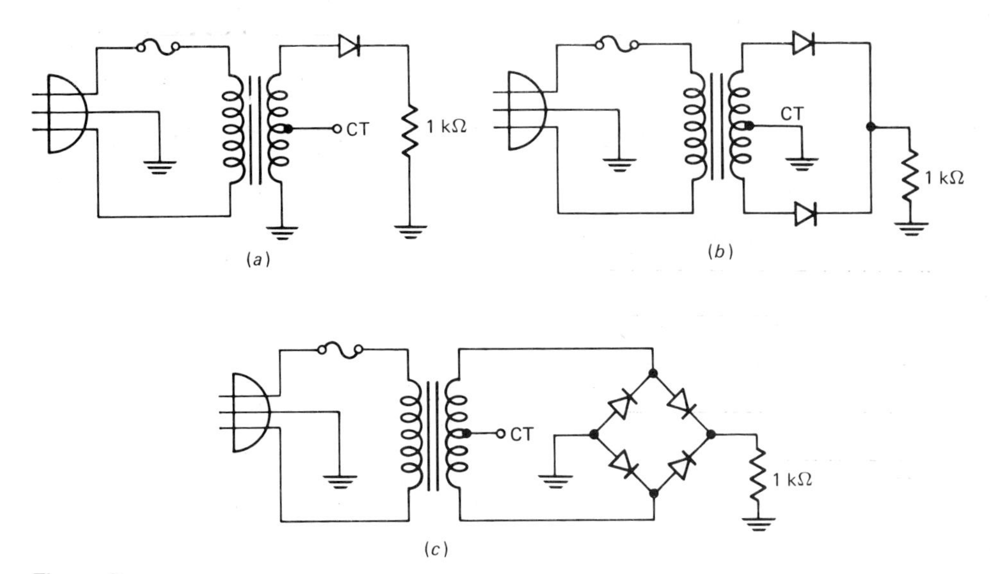

Figure 6

2. Measure the rms voltage V_2 across the secondary winding and record in Table 1.
3. Use the oscilloscope to look at the rectified voltage across the 1-kΩ load resistor. Record the peak voltage V_p and period T of the half-wave signal (Table 1).
4. Measure the dc voltage across the load. Record the value V_{DC} in Table 1.
5. Connect the center-tap rectifier of Fig. 6*b*.
6. Measure and record the rms voltage across the upper half of the secondary winding (use Table 2).
7. With the oscilloscope, look at the full-wave signal across the load resistor. Record the peak voltage and period in Table 2.
8. Measure the dc voltage across the load resistor and record in Table 2.
9. Connect the bridge rectifier of Fig. 6*c*.
10. Measure the rms voltage across the secondary winding and record the value in Table 3.
11. Look at the full-wave signal across the load resistor. Record the peak voltage and period in Table 3.
12. Measure and record the dc voltage across the load resistor.

DATA

Table 1. Half-wave Rectifier

V_2	15.16V
V_P	20V
T	16ms
V_{DC}	6.46V

Table 2. Center-tap Rectifier

V_2	7.59V
V_P	10V
T	8ms
V_{DC}	6.18V

Table 3. Bridge Rectifier

V_2	15.11
V_P	20V
T	8ms
V_{DC}	12.28V

7 the capacitor-input filter

REQUIRED READING

Pages 75–82

EQUIPMENT

1 center-tapped transformer, 12.6 V (Triad F-25X or equivalent) with fused line cord
4 silicon diodes: 1N4001 (or equivalent)
3 ½-W resistors: 220 Ω, 1 kΩ, 10 kΩ
2 capacitors: 1 μF and 470 μF (25-V rating or better)
1 VOM (or DVM if available)
1 oscilloscope

PROCEDURE

1. Measure the resistances of the primary and secondary windings. Record the values in Table 1.
2. Build the circuit of Fig. 7, using $R_{SURGE} = 0$, $C = 0$, and $R_L = 10$ kΩ.

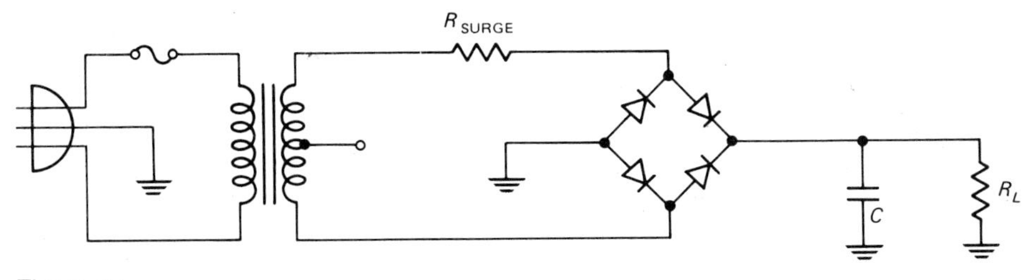

Figure 7

3. Measure and record the dc load voltage (Table 2). With a dc-coupled oscilloscope, look at the voltage waveform across the load resistor. Sketch the waveform here, indicating voltage levels:

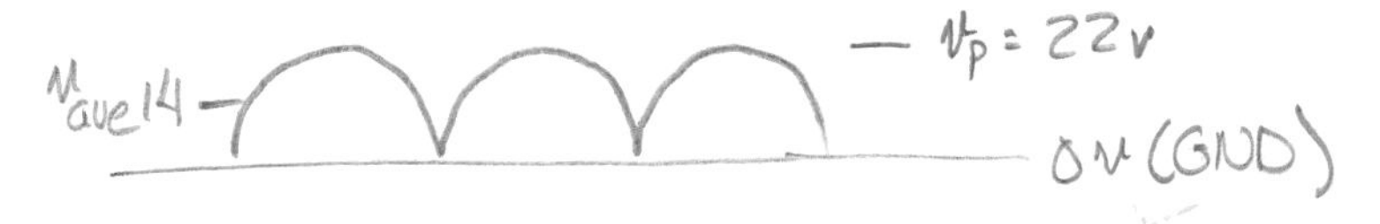

4. Use $C = 1\ \mu F$. Measure and record the dc load voltage (Table 2). Sketch the load-voltage waveform here:

5. Change C to $470\ \mu F$. Measure and record the dc load voltage (Table 2). Sketch the waveform here:

6. Switch the oscilloscope from dc input to ac input. Increase the sensitivity until you can see the ripple. Record the peak-to-peak ripple in Table 2.
7. Use $R_{SURGE} = 220\ \Omega$ and $C = 1\ \mu F$. Measure and record the dc load voltage in Table 3. With a dc-coupled oscilloscope, look at the load voltage. Sketch the waveform here:

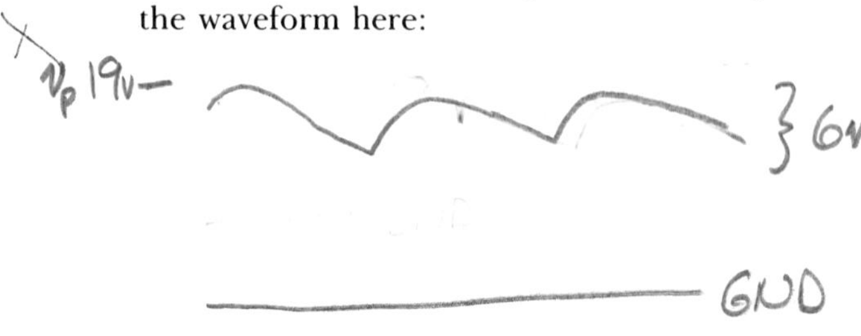

8. Change R_L to 1 kΩ. Measure and record the dc load voltage in Table 3. Sketch the load-voltage waveform here:

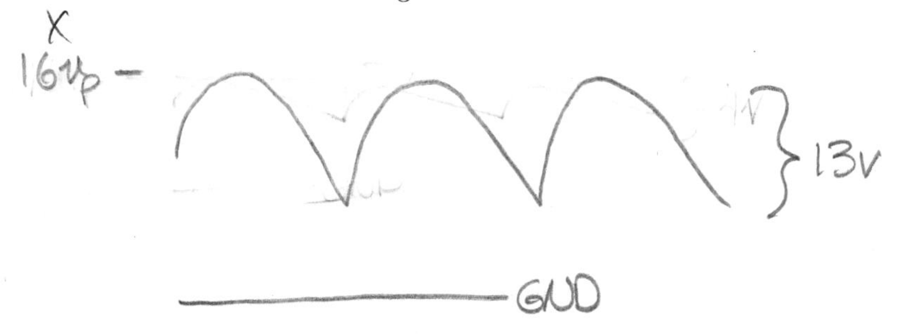

DATA

Table 1. Transformer Resistances

R_1	1.2Ω
R_2	43Ω

Table 2. $R_{SURGE} = 0$ **and** $R_L = 10\ k\Omega$

V_{DC} for step 3	12.6V
V_{DC} for step 4	17.3V
V_{DC} for step 5	20.0 V
$V_{rip(pp)}$	30mV

Table 3. $R_{SURGE} = 220\ \Omega$ **and** $C = 1\ \mu F$

V_{DC} for step 7	16.6V
V_{DC} for step 8	10.5V

8 *voltage multipliers*

REQUIRED READING

Pages 86–88

EQUIPMENT

1 center-tapped transformer, 12.6 V (Triad F-25X or equivalent) with fused line cord
3 silicon diodes: 1N4001 (or equivalent)
1 ½-W resistor: 6.8 kΩ
3 capacitors: 47 μF (25-V rating or better)
1 VOM (or DVM if available)
1 oscilloscope

PROCEDURE

1. Build the circuit of Fig. 8.

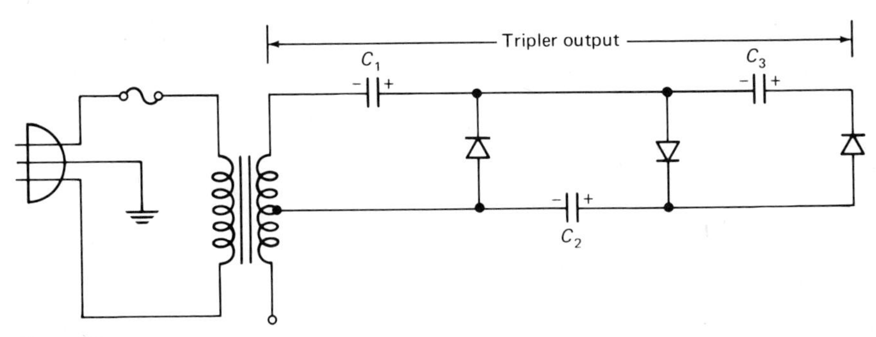

Figure 8

2. Measure and record the rms voltage across half the secondary winding (Table 1).
3. Calculate the peak voltage across half the secondary winding and record the value in Table 1.

4. With a dc-coupled oscilloscope, look at the voltage across C_1. Use the VOM (or DVM) to measure the dc voltage across C_1 and record its value in Table 1 under V_{NL}.
5. Connect a 6.8-kΩ resistor across C_1. Measure and record the dc voltage in Table 1 under V_{FL}.
6. Remove the 6.8-kΩ load resistor.
7. With the oscilloscope, look at the voltage across C_2. Measure and record the dc voltage across C_2 (use V_{NL} of Table 2).
8. Connect the 6.8-kΩ resistor across C_2. Measure and record the dc voltage across C_2 (use V_{FL} of Table 2).
9. Remove the 6.8-kΩ resistor.
10. With the oscilloscope, look at the voltage across the tripler output. Measure and record the dc voltage across the tripler output (use V_{NL} of Table 3).
11. Connect the 6.8-kΩ resistor across the tripler output. Measure and record the dc voltage across the tripler output (use V_{FL} of Table 3).
12. Use the VOM (or DVM) to measure the approximate ripple voltage across the tripler output and record the value in Table 3. (This is approximate because most voltmeters are calibrated to read the rms value of a sine wave. The ripple is not a sine wave; therefore, the reading is an approximation.) SMALLER VALUES of C HAVE LOWER VR

DATA

Table 1. ×1 Output

V_{rms}	7.49V
V_P	10.60V
10.3 V_{NL}	10.16V_{DC}
9.7 V_{FL}	9.72V_{DC}

VR = 4.5%

Table 2. Doubler Output

20.6 V_{NL}	20.3V_{DC}
17.8 V_{FL}	18.6V_{DC}

VR = 9%

Table 3. Tripler Output

30.9 V_{NL}	30.4 V_{DC}
27.4 V_{FL}	25.4 V_{DC}
V_r	2.12V

VR = 19%

R_L = 12k
V_{NL} 30.5V
V_{FL} 27.3V
VR 12%

R_L = 24k
V_{NL} = 30.8V
V_{FL} = 28.9V
VR = 6.6%

R_L = 36k
V_{NL} 30.8V
V_{FL} 29.4V
VR 4.8%

9 *the zener regulator*

REQUIRED READING

Pages 91–94

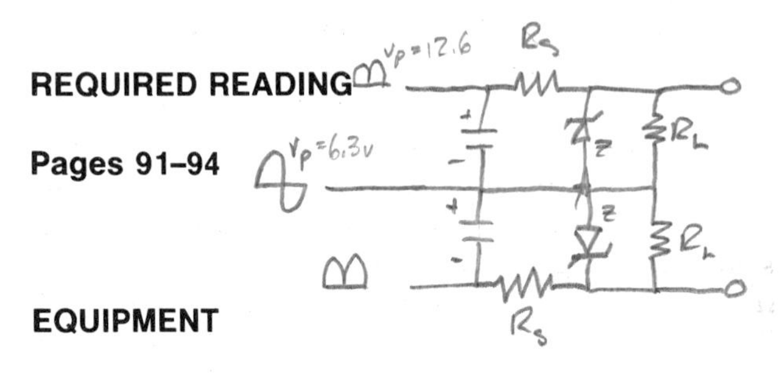

EQUIPMENT

1 center-tapped transformer, 12.6 V (Triad F-25X or equivalent) with fused line cord
4 silicon diodes: 1N4001 (or equivalent)
2 zener diodes: 1N753
4 ½-W resistors: two 150 Ω, two 470 Ω
2 capacitors: 470 μF (25-V rating or better)
1 VOM (or DVM if available)
1 oscilloscope

PROCEDURE

1. Build the split-supply of Fig. 9.

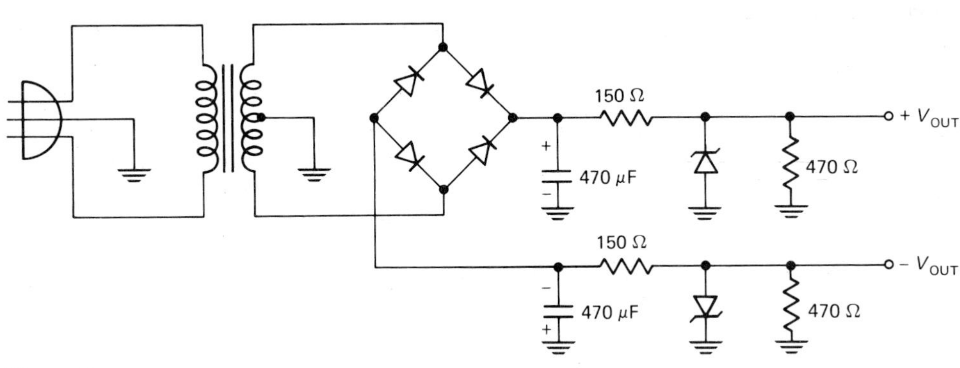

Figure 9

2. Measure the unregulated positive dc voltage across the upper filter capacitor. Record the value in Table 1 as $+V_{IN}$.
3. Measure and record the regulated positive output voltage (Table 1, $+V_{OUT}$).
4. Measure the unregulated negative input voltage. Record in Table 2 as $-V_{IN}$.
5. Measure and record the regulated negative output voltage (Table 2, $-V_{OUT}$).
6. With the ac input of the oscilloscope, look at the ripple across the upper filter capacitor. Record the peak-to-peak ripple in Table 3 as $V_{r(\mathrm{IN})}$.
7. Look at the ripple across the upper zener diode and record the peak-to-peak value in Table 3 as $V_{r(\mathrm{OUT})}$.
8. Look at the ripple across the lower filter capacitor. This ripple should be inverted compared to $V_{r(\mathrm{IN})}$ but have approximately the same peak-to-peak value. Likewise, look at the ripple across the lower zener diode which should have approximately the same peak-to-peak value as $V_{r(\mathrm{OUT})}$.
9. Remove the upper 470-Ω load resistor. Measure and record the regulated positive output voltage (Table 3, V_{NL}).
10. Remove the upper zener diode and measure the new value of $+V_{\mathrm{OUT}}$. Record in Table 3.

DATA

Table 1. Positive Regulator

$+V_{\mathrm{IN}}$	9.63
$+V_{\mathrm{OUT}}$	6.27

Table 2. Negative Regulator

$-V_{\mathrm{IN}}$	9.66
$-V_{\mathrm{OUT}}$	6.31

Table 3. Ripple and Other Voltages

$V_{r(\mathrm{IN})}$	.3v
$V_{r(\mathrm{OUT})}$	4mV
V_{NL}	6.32v
$+V_{\mathrm{OUT}}$	10.25v

10 clippers and peak rectifiers

REQUIRED READING

Pages 94–96

EQUIPMENT

1 audio generator
1 power supply: adjustable from at least 1 to 15 V
2 diodes: 1N914 (or almost any small-signal silicon diode)
4 ½-W resistors: 470 Ω, 1 kΩ, 10 kΩ, 100 kΩ
1 capacitor: 1 μF (10-V rating or better)
1 VOM
1 oscilloscope

PROCEDURE

1. Connect the positive clipper of Fig. 10*a*. (The 1 kΩ is a dc return.)
2. Adjust the source to get 1 kHz and 20 V peak-to-peak across the input. (If the source cannot put out 20 V, set it to maximum.) Record V_{pp} at the top of Table 1.
3. Move oscilloscope leads to output. You should get a positively clipped sine wave. Record the positive and negative peak values in Table 1. (You must use the dc input of the oscilloscope.)
4. Reverse the polarity of the diode and look at the output waveform. It should be negatively clipped. Record the positive and negative peak values in Table 1.
5. Connect the combination clipper of Fig. 10*b*.
6. Look at the output waveform. Measure and record the positive and negative peaks in Table 1.
7. Connect the variable clipper of Fig. 10*c*.
8. Look at the output with an oscilloscope (dc input). When you vary the dc source, the positive clipping level should vary from a low value to a high value. If it does, write "yes" under positive peak in Table 1. Measure and record the negative peak.

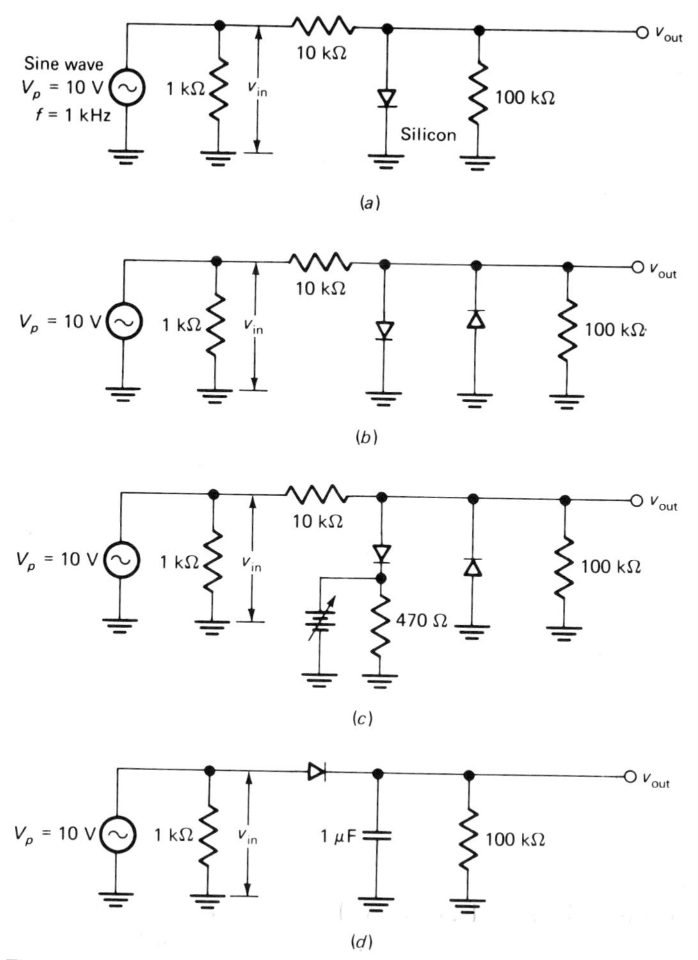

Figure 10

9. Connect the peak rectifier of Fig. 10*d*.
10. Adjust the source to get 1 kHz and 10 V peak across the input. If the signal source won't put out 10 V peak, set it at maximum. Record V_p in Table 2.
11. Look at the output voltage with the oscilloscope. It should be a dc voltage approximately equal to V_p in Table 2. Record this as $V_{dc(scope)}$ in Table 2.
12. Switch to ac input on the oscilloscope and increase sensitivity until you can measure the ripple accurately. Record the peak-to-peak ripple in Table 2.
13. Use the VOM to measure the dc output voltage in Fig. 10*d*. Record this as $V_{dc(VOM)}$ in Table 2.

14. Because the VOM has input resistance on its voltmeter ranges, it changes the resistance across the 1-μF capacitor. While looking at the output ripple on the oscilloscope, connect and disconnect the VOM. What happens to the ripple when the VOM is disconnected? Record "bigger" or "smaller" in Table 2.

DATA

Table 1. Clippers: V_{pp} = 20v

	pos peak	neg peak
pos clipper	0v	9.5v
neg clipper	9.5v	.5v
combo clipper	.6v	.6v
variable clipper	YES	.6v

Table 2. Peak Rectifier

V_p	10v
$V_{dc(scope)}$	9.5v
v_{rip}	100mV
$V_{dc(VOM)}$	9.31v
ripple change	NO CHANGE (DMM)

11 clamping action

REQUIRED READING

Pages 97–104

EQUIPMENT

1 audio generator
2 diodes: 1N914 (or almost any small-signal silicon diode)
3 ½-W resistors: 1 kΩ, 10 kΩ, 100 kΩ
2 capacitors: 1 μF each (20-V rating or better)
1 VOM
1 oscilloscope

PROCEDURE

1. Connect the positive clamper of Fig. 11*a*.
2. Adjust the source to get 1 kHz and 20 V peak-to-peak across the input (or nearest possible). Record V_{pp} in Table 1.
3. With oscilloscope on dc input, look at the output of Fig. 11*a*. It should be a positively clamped sine wave. Measure and record the positive and negative peaks (Table 1).
4. Keep the oscilloscope on the output and vary the input voltage. Notice how the negative peak is clamped near zero while the positive peak moves up and down.
5. Connect the peak-to-peak detector of Fig. 11*b*. Adjust the source to get 1 kHz and 20 V peak-to-peak across the input (or nearest possible).
6. Look at the voltage across the first diode. It should be a positively clamped signal.
7. Look at the output. It should be a dc voltage with a small ripple. Switch the oscilloscope to ac input and high sensitivity to measure the ripple. Record v_{rip} in Table 2.
8. Measure the dc output voltage in Fig. 11*b* with the VOM. Record V_{DC} in Table 2.

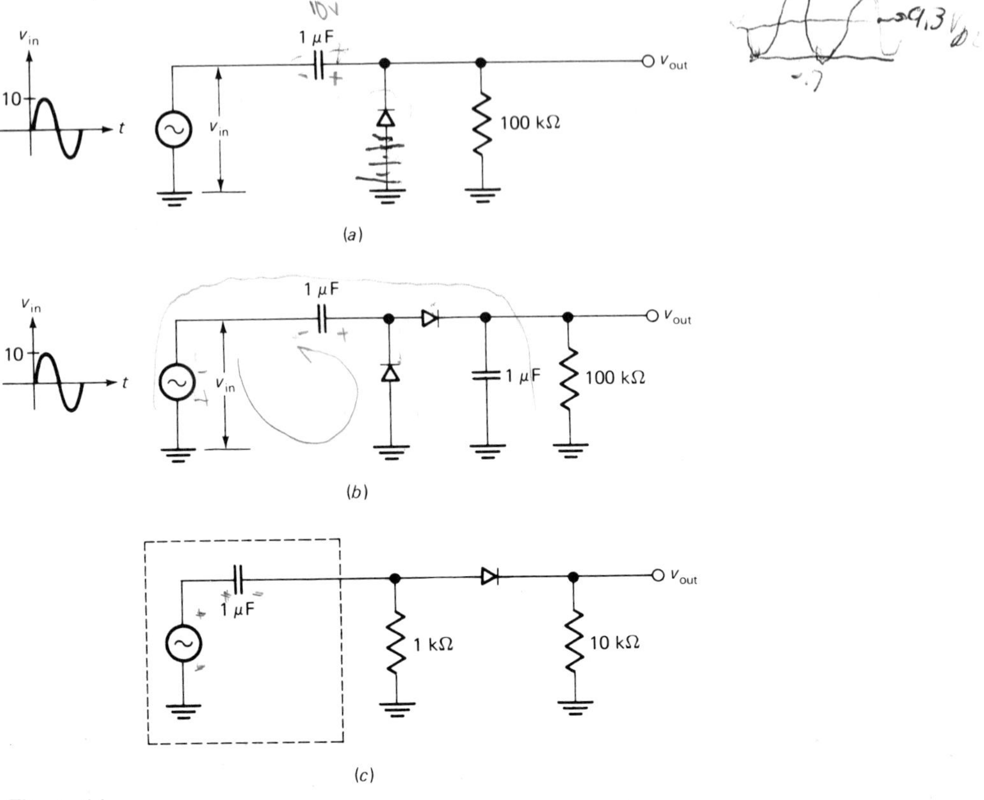

Figure 11

9. Connect the circuit of Fig. 11*c*. The inside of the dashed box simulates a capacitively coupled source. The 1-kΩ resistor is a dc return.
10. Adjust the source to get 1 kHz and 20 V peak-to-peak (or nearest possible) across the 1-kΩ resistor.
11. Look at the output with the oscilloscope. It should be a half-wave signal. Record the peak value in Table 3.
12. Disconnect the dc return. Record the output peak value in Table 3.

DATA

Table 1. Positive Clamper

V_{pp}	20v
out pos peak	20v
out neg peak	0v

Table 2. Peak-to-Peak Detector

v_{rip}	150mV
V_{DC}	17.54v

Table 3. Capacitively Coupled Source

V_{peak} (with dc return)	9v
V_{peak} (without dc return)	120mV

12 *the ce connection*

REQUIRED READING

Pages 121–125

EQUIPMENT

2 power supplies: 9 V
2 ½-W resistors: 100 Ω, 100 kΩ
3 transistors: 2N3904 (or almost any small-signal *npn* silicon transistor)
1 VOM

PROCEDURE

1. Connect the circuit of Fig. 12, using one of the transistors. Measure and record V_{BB} and V_{CC} (top of Table 1).

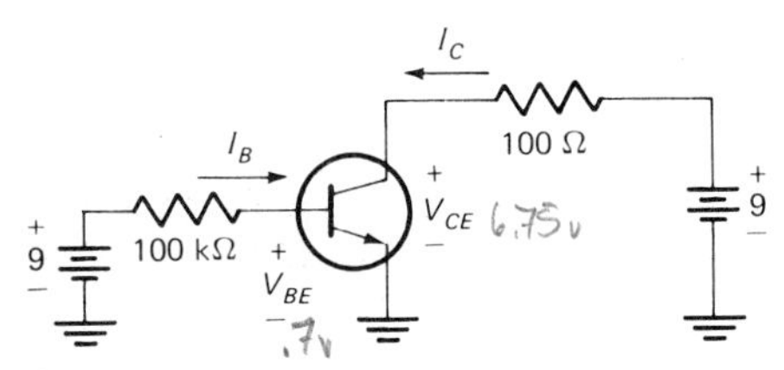

Figure 12

2. Measure and record the values of V_{BE} and V_{CE}.
3. Use the VOM as an ammeter to measure I_C and I_B. Record these values in Table 1.
4. Calculate the values of V_{CB}, I_E, α_{dc}, and β_{dc} in Fig. 12. Record in Table 2.
5. Connect the circuit of Fig. 12 using another transistor. Repeat steps 2 through 4.
6. Connect the circuit of Fig. 12 using a third transistor. Repeat steps 2 through 4.

DATA

Table 1. V_{BB} = 9V ; V_{CC} = 9V

Test	V_{BE}	V_{CE}	I_C	I_B
1	.70v	6.75v	22.2mA	.084mA
2	.68v	7.85v	11.68mA	.085mA
3	.68v	8.00v	10.3mA	.085mA

Table 2. Calculations

Test	V_{CB}	I_E	α_{dc}	β_{dc}
1				
2				
3				

TEST	V_{CB}	I_E	α_{dc}	β_{dc}
1	8.59v	22.284mA	.996	264
2	9.23v	11.765mA	.992	137
3	9.17v	10.385mA	.992	121

13 collector curves

REQUIRED READING

Pages 125–130

EQUIPMENT

1 power supply: 9 V
1 power supply: adjustable from at least to 1 to 10 V
1 transistor: 2N3904 (or almost any small-signal *npn* silicon transistor)
2 ½-W resistors: 100 Ω, 100 kΩ
1 decade resistance box (or substitute 220 kΩ, 470 kΩ, 1 MΩ)
1 oscilloscope

PROCEDURE

1. Set up the oscilloscope as follows:
   a. Horizontal sensitivity to 1 V/cm (dc input).
   b. Vertical sensitivity to 0.1 V/cm (dc input).
   c. Center the undeflected spot in the upper lefthand corner of the screen.
2. Connect the circuit of Fig. 13*a*. Notice the collector current passes through the 100-Ω resistor. Because of this, the voltage to the vertical input is proportional to I_C. In fact, each milliampere of collector current produces 100 mV of vertical input.
3. Set the resistance box R to 1 MΩ.
4. Vary the V_{CC} supply back and forth rapidly between minimum and 10 V. You will see the first collector curve of Fig. 13*b*.
5. Change R to 470 kΩ. Again vary the V_{CC} supply rapidly between low and high voltage. You should see the second curve of Fig. 13*b*.
6. Change R to 220 kΩ and vary V_{CC} rapidly. You should see a third collector curve.
7. Return R to 1 MΩ. Set the V_{CC} supply to produce each V_{CE} listed in Table 1. For each V_{CE}, read and record the value of I_C.
8. In a similar way, fill in the rest of Table 1 for the other values of R.
9. This step and the next two are optional; ask your instructor. If an audio

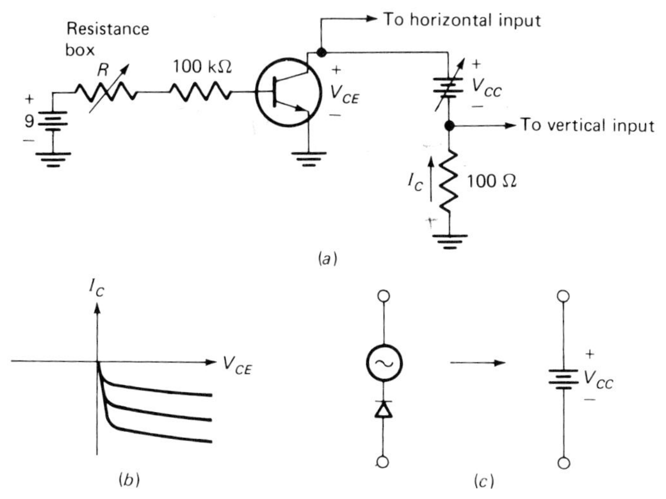

Figure 13

generator with a *floating output* is available, connect a diode in series with it as shown in Fig. 13*c*. Then substitute this generator-diode combination for the V_{CC} supply of Fig. 13*a*.

10. Adjust the audio generator to 100 Hz and enough signal to produce a collector curve. Each time you change the resistor substitution box, you get a different collector curve.

11. If a transistor curve tracer is available, use it to look at the collector curves of your transistor.

I_B @ 1.1 MEG $= \frac{8.3}{1.1\text{ MEG}} = 7.55\,\mu A$

$\frac{4.3}{7.55\,\mu A} = 570k$ (560K)

DATA

Table 1. Collector Curves

	V_{CE}	0	2	4	6	8	10
$R =$ 1 MΩ (560K)	I_C	0mA	3mA	3.2mA	3.5mA	3.8mA	4mA
$R =$ 470 kΩ	I_C	0mA	5.6mA	6.2mA	6.6mA	7.0mA	7.4mA
$R =$ 220 kΩ	I_C	0mA	8.8mA	10.8mA	12mA	13mA	14.4mA

I_B @ 470k $= \frac{8.3}{570K} = 14.6\,\mu A$ $\qquad \frac{4.3}{14.6\,\mu A} = R = 295.5\,k\Omega$ (300 kΩ)

I_B @ 220k $= \frac{8.3}{320K} = 25.9\,\mu A$ $\qquad \frac{4.3}{25.9\,\mu A} = R = 166.0\,k\Omega$ (160 kΩ)

14 *base bias*

THEORY

The Q point in a base-biased circuit is heavily dependent on the value of β_{dc}. One way to desensitize the circuit is to add an emitter resistor as shown in Fig. 14*b*. If we ignore the V_{BE} drop, the ideal value of emitter current is

$$I_E \cong \frac{V_{BB}}{R_E + R_B/\beta_{dc}} \tag{14–1}$$

To a second approximation, we can allow 0.7 V for the V_{BE} drop. In this case,

$$I_E \cong \frac{V_{BB} - 0.7}{R_E + R_B/\beta_{dc}} \tag{14–2}$$

You will use these two formulas in this experiment.

REQUIRED READING

Pages 135–143

EQUIPMENT

1 power supply: 15 V
2 transistors: 2N3904 (or almost any small-signal *npn* silicon transistor)
5 ½-W resistors: 220 Ω, 470 Ω, 2.2 kΩ, 22 kΩ, 100 kΩ
1 decade resistance box (or substitute a 1-MΩ potentiometer)
1 VOM

PROCEDURE

1. The fixed-base-current circuit of Fig. 14*a* is not a stable biasing circuit, but it is a good way to measure β_{dc}. By a straightforward derivation, when $V_{CE} \cong 1$ V,

$$\beta_{dc} \cong \frac{R_B}{R_C} \tag{14–3}$$

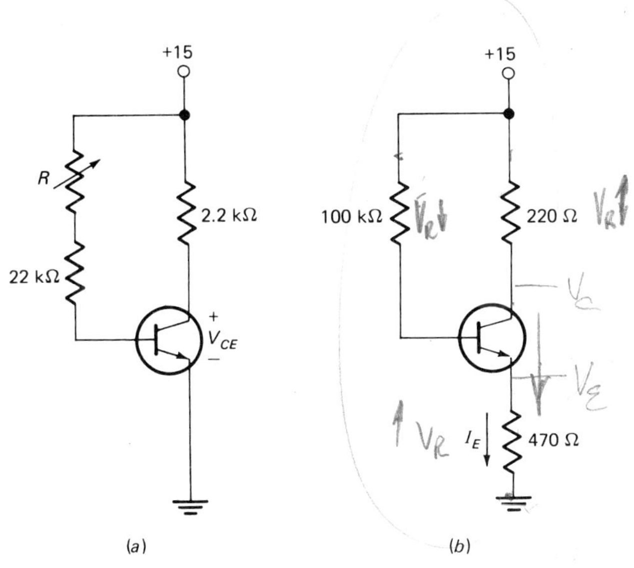

Figure 14

$$I_C \cong \frac{V_{CC}}{R_C} \tag{14–4}$$

In other words, you adjust R in Fig. 14*a* until V_{CE} is approximately 1 V; then, you can use Eqs. (14–3) and (14–4) to calculate β_{dc} and I_C.

2. Connect the circuit of Fig. 14*a* using one of the transistors.
3. Adjust R to get a V_{CE} of 1 V. Record the value of R in Table 1. (If R is a potentiometer instead of a decade box, you will have to disconnect it and measure its resistance.) In Fig. 14*a*, notice the total base resistance R_B equals R plus 22 kΩ. Record the value of R_B in Table 1.
4. With Eqs. (14–3) and (14–4), calculate the values of β_{dc} and I_C. Record in Table 1.
5. Repeat steps 2 through 4 for the second transistor.
6. With the values of Table 1, calculate the ideal and second-approximation values of I_E in Fig. 14*b*. Record the I_E values in Table 2.
7. Connect the base-biased circuit of Fig. 14*b*.
8. Use the VOM as an ammeter and measure I_E. Record this current in Table 3.
9. Use the VOM as a voltmeter and measure V_B, V_E, and V_C. Record these values in Table 3. (Remember: a single subscript on voltage means the voltage is with respect to ground.)
10. Calculate and record the value of V_{CE}.
11. Repeat steps 7 through 10 for the second transistor.
12. If a curve tracer or other transistor tester is available, measure the β_{dc} of

each transistor for an I_C of approximately 5 mA. The values should be similar to the β_{dc} values of Table 1.

DATA

Table 1. β_{dc} **Values**

Test	R	R_B	β_{dc}	I_C
1	157k	179k	81.4	6.8mA
2	220k	242k	110	6.8mA

Table 2. Calculations

Test	$I_{E\,(ideal)}$	$I_{E\,(second)}$
1	8.83mA	8.42mA
2	10.9mA	10.4 mA

Table 3. Measurements

Test	I_E	V_B	V_E	V_C	V_{CE}
1	15.7mA	8.01	7.31	11.51	4.07 (4.2)
2	11.4mA	6.10	5.40	12.45	7.20 (7.05)

$V_{CE} = V_C - V_E$

15 voltage-divider and emitter bias

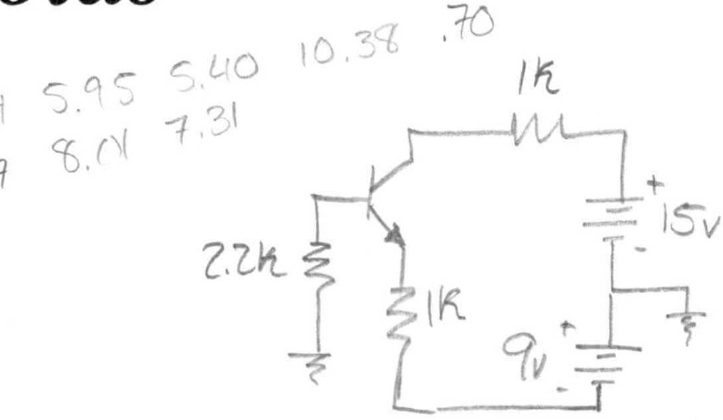

REQUIRED READING

Pages 143–148, 152–155

EQUIPMENT

2 power supplies: 9 V and 15 V
3 transistors: 2N3904 (or almost any small-signal *npn* silicon transistor)
4 ½-W resistors: two 1 kΩ, 2.2 kΩ, 4.7 kΩ
1 VOM

PROCEDURE

1. In Fig. 15*a*, calculate V_B, V_E, I_E, and V_C. Record these values in Table 1.
2. Connect the circuit of Fig. 15*a*. Measure and record the values of V_B, V_E, and V_C (Table 2).

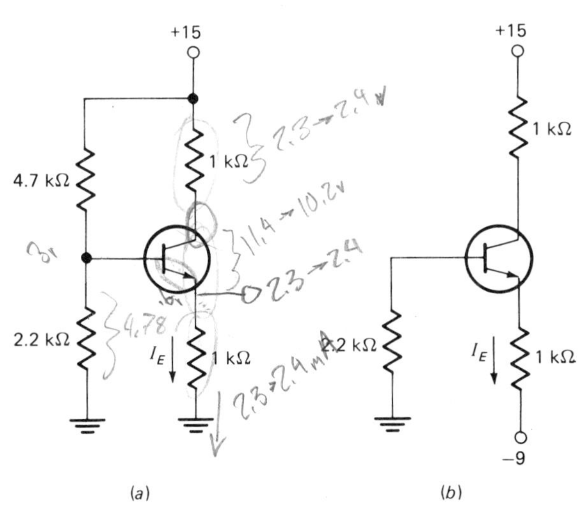

Figure 15

3. Using the VOM as an ammeter, measure and record I_E (Table 2).
4. With the VOM still connected as an ammeter, substitute the two other transistors and notice how I_E stays almost constant.
5. In Fig. 15*b*, calculate the ideal and second-approximation values of I_E. [Second approximation: $I_E \cong (V_{EE} - 0.7)/R_E$.] Record in Table 3.
6. Connect the circuit of Fig. 15*b*. Measure and record V_B, V_E, and V_C (Table 4).
7. Using the VOM as an ammeter, measure I_E and record the value in Table 4.
8. With the VOM still connected as an ammeter, substitute the two other transistors and notice how I_E remains almost constant.

DATA

Table 1.

V_B	4.78v
V_E	4.08
V_C	10.92v
I_E	4.08 mA

Table 2.

V_B	4.84 v
V_E	4.16v
V_C	11.07v
I_E	4.06mA

Table 3.

$I_{E\,\text{(ideal)}}$	9mA
$I_{E\,\text{(second)}}$	8.3mA

Table 4.

V_B	-.16
V_E	-.84
V_C	6.96
I_E	8.04mA

V_{CE} = 7.84

16 collector-feedback bias

REQUIRED READING

Pages 148–152

EQUIPMENT

1 power supply: 15 V
3 transistors: 2N3904 (or almost any small-signal *npn* silicon transistor)
2 ½-W resistors: 1 kΩ, 100 kΩ
1 VOM

PROCEDURE

1. In Fig. 16, use a β_{dc} of 25 and Eq. (7–19) to calculate I_E and V_C. Record these values in Table 1.

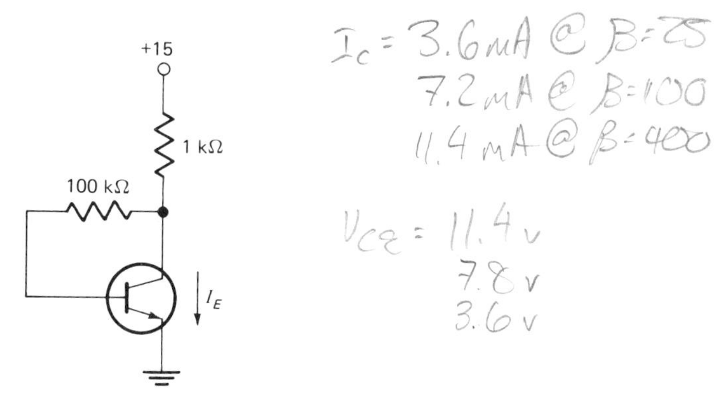

Figure 16

2. Repeat the calculations for a β_{dc} of 100 and a β_{dc} of 400. Record values in Table 1.
3. Connect the circuit of Fig. 16.
4. Measure and record the value of I_E (Table 2).

5. Measure and record the value of V_C (Table 2).
6. Repeat steps 3 through 5 for the two other transistors.
7. If a curve tracer or other transistor tester is available, measure the β_{dc} of each transistor for an I_C of approximately 7 mA. Record the values in Table 2. (If you have no equipment for measuring β_{dc}, leave the β_{dc} spaces blank in Table 2.)

DATA

Table 1. Calculations

12.14
7.85
3.6

β_{dc}	I_E	V_C
25	2.86mA	12.14v
100	7.15mA	7.85v
400	11.4mA	3.6v

Table 2. Measurements

Test	I_E	V_C	β_{dc}
1	11.26mA	3.77	327
2	9.93mA	5.14v	200
3	7.83mA	7.17v	46

7.83mA 7.17v

17 *upside-down* pnp *biasing*

REQUIRED READING

Pages 155–160

EQUIPMENT

1 power supply: 15 V
1 transistor: 2N3906 (or almost any small-signal *pnp* silicon transistor)
5 ½-W resistors: two 1 kΩ, 2.2 kΩ, 4.7 kΩ, 100 kΩ
1 VOM

PROCEDURE

1. In Fig. 17*a*, use Eq. (7–9) and calculate the values of V_B, V_E, I_E, and V_C. Record in Table 1.
2. Connect the circuit of Fig. 17*a*. Measure and record V_B, V_E, and V_C (Table 2).

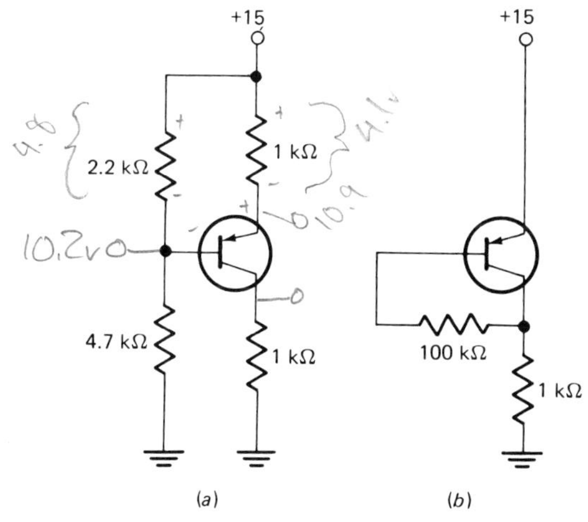

Figure 17

3. Use the VOM as an ammeter and measure I_E. Record this value in Table 2.
4. In Fig. 17*b*, use a β_{dc} of 100 and Eq. (7–19) to calculate I_E and V_C. Record these values in Table 3.
5. Connect the circuit of Fig. 17 *b*. Measure and record I_E and V_C (Table 4).
6. If a curve tracer or other transistor tester is available, measure β_{dc}. Record the value in Table 4.

DATA

Table 1.

V_B	10.2V
V_E	10.9V
I_E	14.9mA
V_C	4.1V

Table 2.

V_B	10.16V
V_E	10.81V
I_E	4.20V
V_C	4.15V

Table 3.

β_{dc}	100
I_E	7.15mA
V_C	7.15V

Table 4.

β_{dc}	160
I_E	9.08mA
V_C	9.08V

18 coupling and bypass capacitors

REQUIRED READING

Pages 170–174

EQUIPMENT

1 audio generator
4 ½-W resistors: two 22 kΩ, 68 kΩ, 100 kΩ
1 capacitor: 0.022 μF
1 oscilloscope

PROCEDURE

1. Use Eq. (8–2) to calculate the lowest frequency that is well coupled in Fig. 18*a*. Fill in the values of f_{low}, $10 f_{\text{low}}$, and $0.1 f_{\text{low}}$ in Table 1 under f.
2. Connect the circuit of Fig. 18*a*.

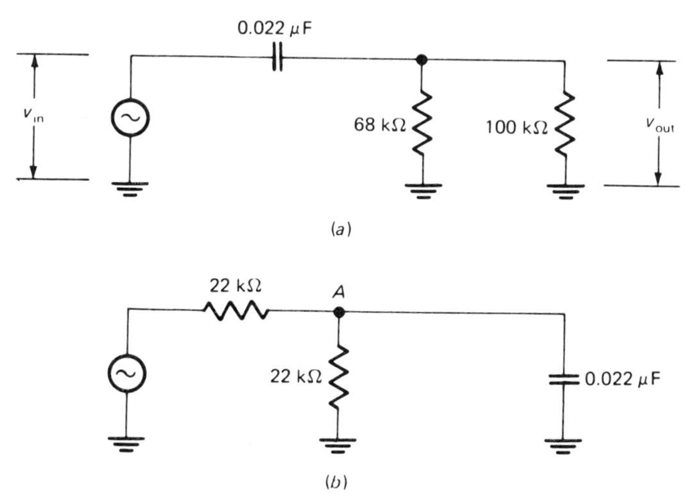

Figure 18

3. Adjust the audio generator to get a frequency of f_{low} and an input voltage v_{in} of 1 V peak-to-peak on the oscilloscope.
4. Measure the output voltage v_{out} and record in Table 1.
5. Change the frequency to 10 f_{low} and readjust to get a v_{in} of 1 V pp. Measure v_{out} and record (Table 1).
6. Change the frequency to 0.1 f_{low} and check that v_{in} is 1 V pp. Measure and record v_{out}.
7. Use Eq. (8–2) to calculate the lowest frequency that is well bypassed in Fig. 18 *b*. Fill in the values of f_{low}, 10 f_{low}, and 0.1 f_{low} in Table 2.
8. Connect the circuit of Fig. 18 *b* without the capacitor. Adjust the frequency to f_{low}.
9. Set the signal level to 1 V pp across the lower 22-kΩ resistor.
10. Connect the capacitor between point *A* and ground. Then measure and record v_A.
11. Remove the capacitor and change the frequency to 10 f_{low}. Then repeat steps 9 and 10.
12. Remove the capacitor and change the frequency to 0.1 f_{low}. Then repeat steps 9 and 10.

DATA

Table 1. Coupling Capacitor

Test	f	v_{out}
f_{low}	1.14 kHz	1V
10 f_{low}	11.4 kHz	1V
0.1 f_{low}	114 Hz	.5V

Table 2. Bypass Capacitor

Test	f	v_A
f_{low}	4.13 kHz	160 mV
10 f_{low}	41.3 kHz	16 mV
0.1 f_{low}	413 Hz	.8V

19 ac emitter resistance

REQUIRED READING

Pages 181–184

EQUIPMENT

1 audio generator
1 power supply: 9 V
1 power supply: adjustable from at least 1 to 12 V
3 transistors: 2N3904 (or almost any small-signal *npn* silicon transistor)
3 ½-W resistors: 1 kΩ, two 10 kΩ
1 capacitor: 0.1 μF
1 oscilloscope
1 ac millivoltmeter

PROCEDURE

1. Figure 19*a* shows a circuit that measures r_e'. In the dc equivalent circuit of Fig. 19*b*, calculate the value of I_E for each value of $-V_{EE}$ shown in Table 1. Record the values of I_E.
2. Using Eq. (8–5), calculate and record the value of r_e' for each I_E.
3. Connect the circuit of Fig. 19*a*. Set $-V_{EE}$ at -10.7 V. Adjust the audio generator to 1 kHz. With the ac millivoltmeter in parallel with the audio generator, adjust the signal level to 1 V rms.
4. Once the audio generator has been set to 1 V rms, the approximate ac equivalent circuit looks like Fig. 19*c*. Notice most of the 1 V will be across the 10-kΩ resistance because r_e' is very small by comparison (around 25 Ω). This is why the ac current i_e is approximately 0.1 mA as shown.
5. Connect the ac millivoltmeter and the oscilloscope to measure v_{be}, the ac voltage between the emitter and ground in Fig. 19*a*.
6. The oscilloscope should display a small sine wave. If so, read the value of v_{be} with the ac millivoltmeter. Record in Table 2.
7. Reduce the $-V_{EE}$ supply to -5.7 V. The sine wave on the oscilloscope

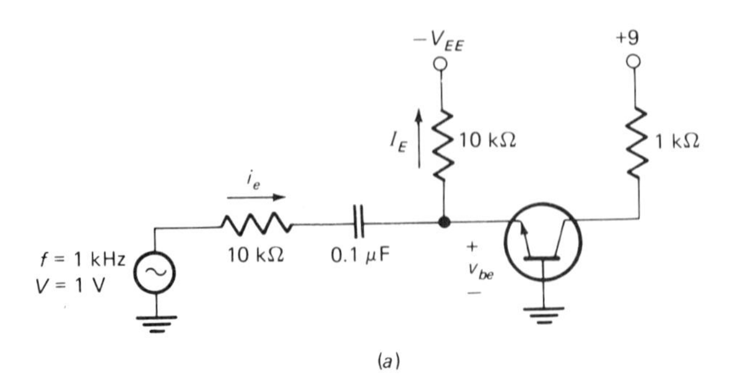

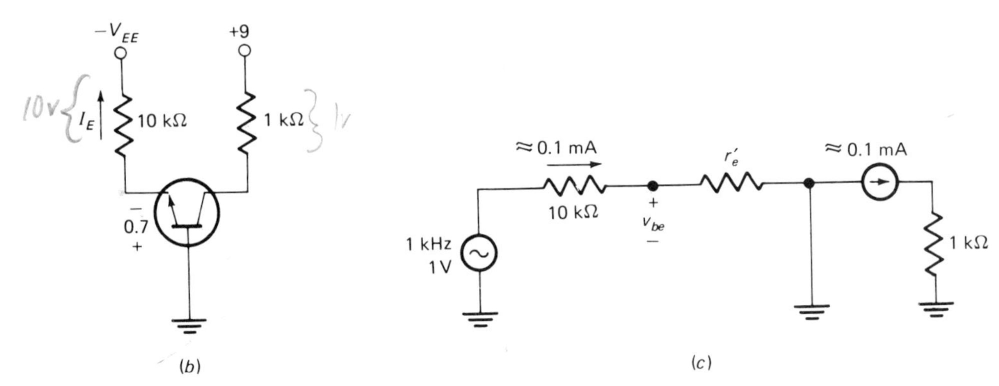

Figure 19

should increase. Read the new value of v_{be} on the ac millivoltmeter and record in Table 2.

8. Reduce the $-V_{EE}$ supply to −3.2 V. Read the new value of v_{be} and record.
9. Since i_e equals 0.1 mA, you can calculate r'_e using Eq. (8–4). Record the calculated values in Table 2.
10. Touch the transistor case and notice it is cool.
11. Return $-V_{EE}$ to −10.7 V and repeat step 6 for the two other transistors, except for recording the values of v_{be} in Table 3.

DATA

Table 1.

$-V_{EE}$	I_E	r'_e
−10.7	1mA	25
−5.7	.5mA	50
−3.2	.25mA	100

Table 2.

$-V_{EE}$	v_{be}	r'_e
−10.7	2.5mV	25
−5.7	5mV	50
−3.2	10.3mV	103

Table 3.

$-V_{EE}$	v_{be} (second)	v_{be} (third)
−10.7	2.5mV	2.8mV

20 base-driven amplifier

REQUIRED READING

Pages 193–209

EQUIPMENT

1 audio generator
1 power supply: 20 V
1 transistor: 2N3904 (or almost any small-signal *npn* silicon transistor)
8 ½-W resistors: 47 Ω, 220 Ω, two 4.7 kΩ, four 10 kΩ
3 capacitors: two 1 μF, one 100 μF (10-V rating or better)
1 oscilloscope

PROCEDURE

1. Figure 20 shows a circuit similar to Fig. 9–7 in the textbook. In this experiment, you will verify that the waveforms resemble those shown in the textbook.
2. Connect the circuit of Fig. 20.

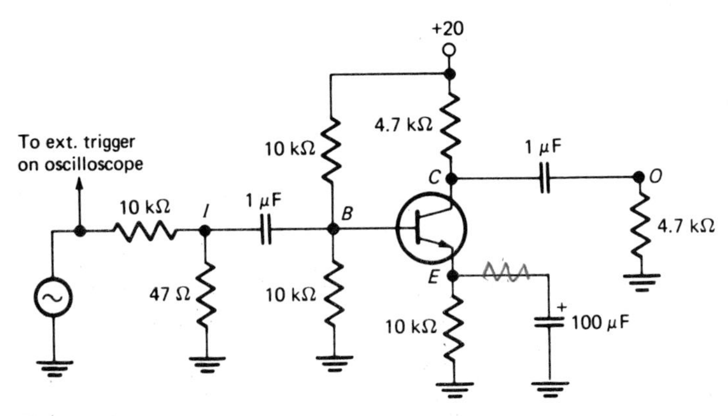

Figure 20

3. The voltage divider at the input is needed to get an adequate trigger for the oscilloscope without overdriving the amplifier. (If necessary, ask your instructor how to set up the external trigger on the oscilloscope.)
4. Adjust the audio generator to 1 kHz. Set the signal level so that a sine wave of 20 mV pp is across the 47-Ω resistor of Fig. 20 (point *I* to ground).
5. Look at the other points in the amplifier (*B, E, C,* and *O*). At each point, use the oscilloscope on dc input to measure the dc voltage; use the oscilloscope on ac input to measure the ac peak-to-peak voltage. Record all voltages in Table 1.
6. Notice the phase inversion of the output signal. That is, the signal at the collector is 180° out of phase with the signal at the base.
7. Swamp the emitter diode by adding a 220-Ω resistor between the emitter and the bypass capacitor.
8. Calculate the swamped voltage gain from base to collector using Eq. (9–3). Record this value under A_{calc} in Table 2.
9. Measure and record ac voltages v_b and v_c.
10. Calculate the voltage gain A using the ratio of measured values v_c and v_b. Record in Table 2.
11. With the oscilloscope across the amplifier output, open the bypass capacitor. Notice how the signal disappears. Increase the sensitivity to maximum and measure the ac collector voltage for this condition. Record in Table 2.

DATA

Table 1.

	point *B*	point *E*	point *C*	point *O*
dc	10.0v	9.3v	15.7v	0v
ac	20mV	0v	1.58v	1.58v

Table 2. $r_e' = 26$

I_E	A_{calc}	v_b	v_c	A	v_c (open bypass)
1 mA	9.75	33mV	.34v	10.3	3mV

21 *input impedance*

REQUIRED READING

Pages 209–217

EQUIPMENT

1 audio generator
2 power supplies: 9 V
3 transistors: 2N3904 (or almost any small-signal *npn* silicon transistor)
5 ½-W resistors: 100 Ω, 220 Ω, 2.2 kΩ, 4.7 kΩ, 10 kΩ
1 decade resistance box (if not available, use a 50-kΩ potentiometer)
2 capacitors: 1 μF, 100 μF (10-V rating or better)
1 ac millivoltmeter

PROCEDURE

1. Calculate r'_e and $z_{in(base)}$ in Fig. 21*a* for each β given in Table 1. Record the values.
2. Connect the circuit of Fig. 21*a*. Adjust the audio generator to 1 kHz. With the ac millivoltmeter, set the signal level across the 100-Ω resistor (point *A*) to 10 mV rms.
3. Adjust the decade resistance box *R* until the ac base voltage (point *B*) drops to 5 mV rms. The value of *R* that produces this base voltage equals the value of $z_{in(base)}$ because the ac voltage across $z_{in(base)}$ is half the driving voltage (see Fig. 21*b*). Record the value of $z_{in(base)}$ in Table 2.
4. Repeat steps 2 and 3 for the two other transistors.
5. Insert a 220-Ω resistor between the emitter and the bypass capacitor (see Fig. 21*c*).
6. Check the signal level to see that it is still 10 mV rms across the 100-Ω resistor (point *A* to ground).
7. Adjust the decade resistance box *R* until the ac signal between the base and ground is 5 mV rms. Record the value of *R* under $z_{in(base)}$ in Table 3.
8. Repeat step 7 for the two other transistors.

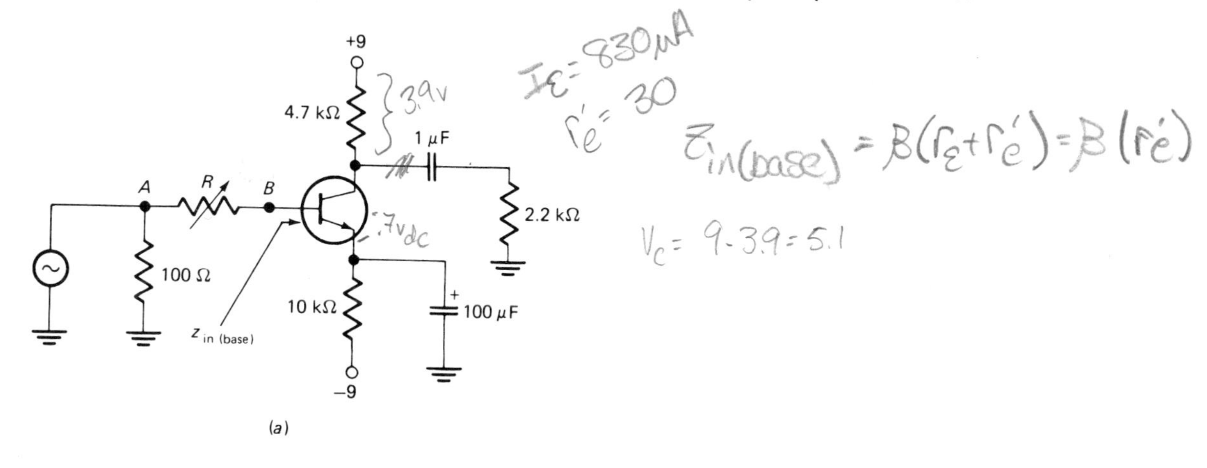

(a)

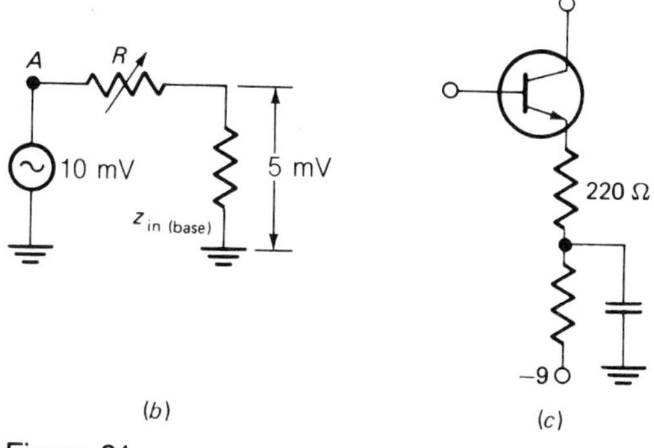

(b)

220 Ω

−9

(c)

Figure 21

DATA

Table 1. Calculations

β	r'_e	$z_{in\,(base)}$
25	30	750
100	30	3000
400	30	12k

Table 2. Without r_E

Test	$z_{in\,(base)}$
1	6.8k
2	3.44k
3	12.9k

Table 3. With r_E of 220 Ω

Test	$z_{in\,(base)}$
1	46k
2	46k
3	100K

22 *the emitter follower*

REQUIRED READING

Pages 219–223

EQUIPMENT

1 audio generator
1 power supply: 20 V
1 transistor: 2N3904 (or similar *npn* silicon transistor)
1 transistor: 2N3906 (or similar *pnp* silicon transistor)
4 ½-W resistors: 100 Ω, 1 kΩ, 10 kΩ, 1 MΩ
1 decade resistance box (if not available, use 50-kΩ potentiometer)
3 capacitors: two 1 μF, 100 μF (10-V rating or better)
1 ac millivoltmeter

PROCEDURE

1. Assume r'_e is 25 Ω in Fig. 22 *a*. With Eq. (9–14 *a*), calculate $z_{\text{in (base)}}$ for each β in Table 1. With Eq. (9–15 *a*), calculate voltage gain A. Record all values in Table 1.
2. Connect the circuit of Fig. 22 *a*. Adjust the audio generator to 1 kHz. With the ac millivoltmeter, set the signal level to 10 mV rms across the input (point A).
3. Adjust the decade resistance box R until v_b is 5 mV rms (point B). The value of R for this condition equals $z_{\text{in (base)}}$ to a close approximation. (The 1-MΩ biasing resistor has only a minor effect.) Record the value of $z_{\text{in (base)}}$ in Table 2.
4. Measure the ac voltage across the 100-Ω resistor (point C). Record in Table 2.
5. Calculate the voltage gain from base to emitter. Record this value of A in Table 2.
6. Connect the upside-down *pnp* emitter follower of Fig. 22 *b*.

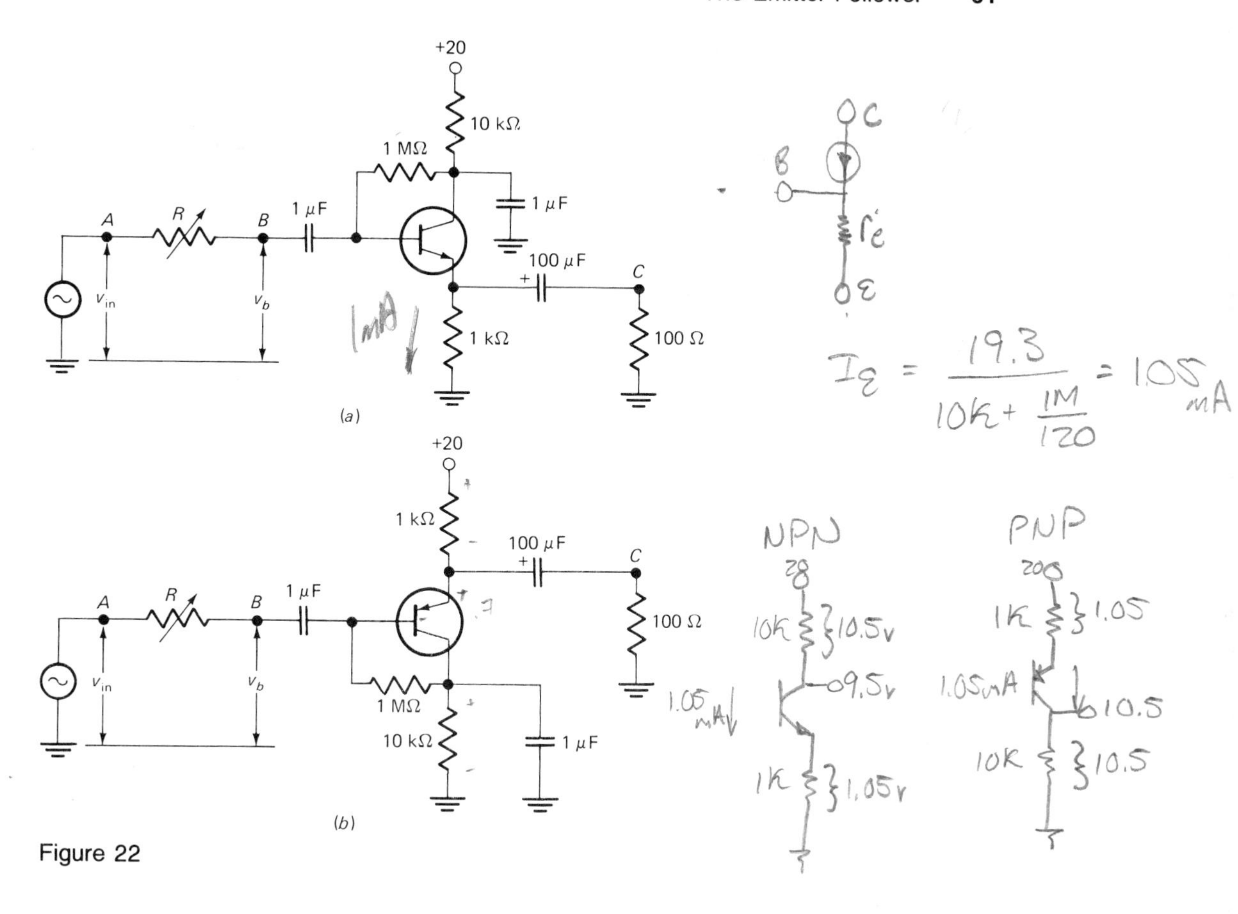

Figure 22

7. Set the frequency to 1 kHz and the signal level to 10 mV rms across the input (point A).
8. Repeat steps 3 through 5.

DATA

Table 1. Calculations

β	$z_{\text{in (base)}}$	A
25	2.9K	.8
100	11.6K	.8
400	46.4K	.8

Table 2. Measurements

Test	$z_{\text{in (base)}}$	v_{out}	A
npn	21.7K	3.9v	.8
pnp	17.8K	3.9v	.8

23 common-base amplifier

REQUIRED READING

Pages 231–233

EQUIPMENT

1 audio generator
2 power supplies: 9 V
1 transistor: 2N3904 (or almost any small-signal *npn* silicon transistor)
4 ½-W resistors: two 1 kΩ, 2.2 kΩ, 4.7 kΩ
2 capacitors: 1 μF
1 oscilloscope

PROCEDURE

1. Analyze the CB amplifier of Fig. 23*a* to get the values of I_E, I_C, V_C, $z_{\text{in(emitter)}}$, r_C, and A (the gain from emitter to collector). Record all values in Table 1.
2. Connect the circuit of Fig. 23*a*.
3. Adjust the audio generator to 1 kHz and the signal level to 1 V pp between point A and ground.
4. Once the audio generator has been set to 1 V pp, the approximate ac equivalent circuit looks like Fig. 23*b*. Because r_e' is much smaller than 1 kΩ, the ac emitter current is approximately 1 mA pp as shown.
5. With the oscilloscope, look at the ac voltage between the emitter and ground. It should be a small sine wave. Measure its peak-to-peak value and record v_e in Table 2.
6. Since i_e is approximately 1 mA (see Fig. 23*b*), you can calculate the value of $z_{\text{in(emitter)}}$ with

$$z_{\text{in(emitter)}} = \frac{v_e}{i_e}$$

   Record this value in Table 2.
7. Look at the ac collector voltage. It should be an amplified sine wave. Measure its peak-to-peak value and record v_c in Table 2.

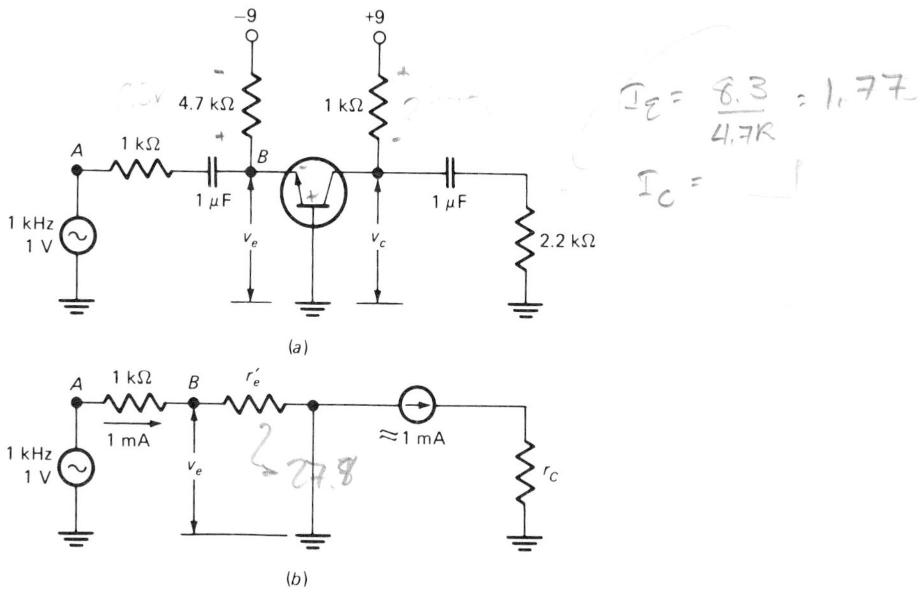

Figure 23

8. Calculate the voltage gain from emitter to collector. Record this value of A in Table 2.

DATA

Table 1. Calculations

I_E	1.77
I_C	1.77
V_C	7.23
$z_{in(emitter)}$	14.1 Ω
r_C	688
A	48.8

Table 2. Measurements

v_e	14.2 mV
$z_{in(emitter)}$	14.3
v_c	.6V
A	46.1

24 *ac load lines*

REQUIRED READING

Pages 244–249

EQUIPMENT

1 audio generator
1 power supply: 9 V
1 transistor: 2N3904 (or almost any small-signal *npn* silicon transistor)
5 ½-W resistors: 100 Ω, 220 Ω, 4.7 kΩ, 10 kΩ, 1 MΩ
1 decade resistance box (if unavailable, use 50-kΩ potentiometer)
2 capacitors: 1μF, 10 μF (10-V rating or better)
1 oscilloscope

PROCEDURE

1. Set up the oscilloscope as follows:
   a. Horizontal sensitivity to 1 V/cm (dc input)
   b. Vertical sensitivity to 100 mV/cm (dc input)
   c. Center spot in upper lefthand corner
2. Connect the circuit of Fig. 24*a*. Notice the collector current passes through the 100-Ω resistor. In fact, each milliampere of collector current produces 100 mV of vertical input to the oscilloscope.
3. With the audio generator turned down to zero, the spot will be deflected. Since vertical sensitivity corresponds to 1 mA/cm and horizontal sensitivity to 1 V/cm, you can measure the values of I_{CQ} and V_{CEQ}. Record these values in Table 1.
4. Adjust the decade resistance box R to 2.5 kΩ. Because the 100-Ω resistor is small, the approximate ac equivalent circuit looks like Fig. 24*b*. Calculate and record the value of r_C (Table 2).
5. Increase the signal level out of the audio generator until the ac load line just hits the saturation and cutoff points. Measure and record the values of $I_{C(sat)}$ and $V_{CE(cutoff)}$ (Table 2).
6. Repeat steps 4 and 5 for the two other values of R shown in Table 2.

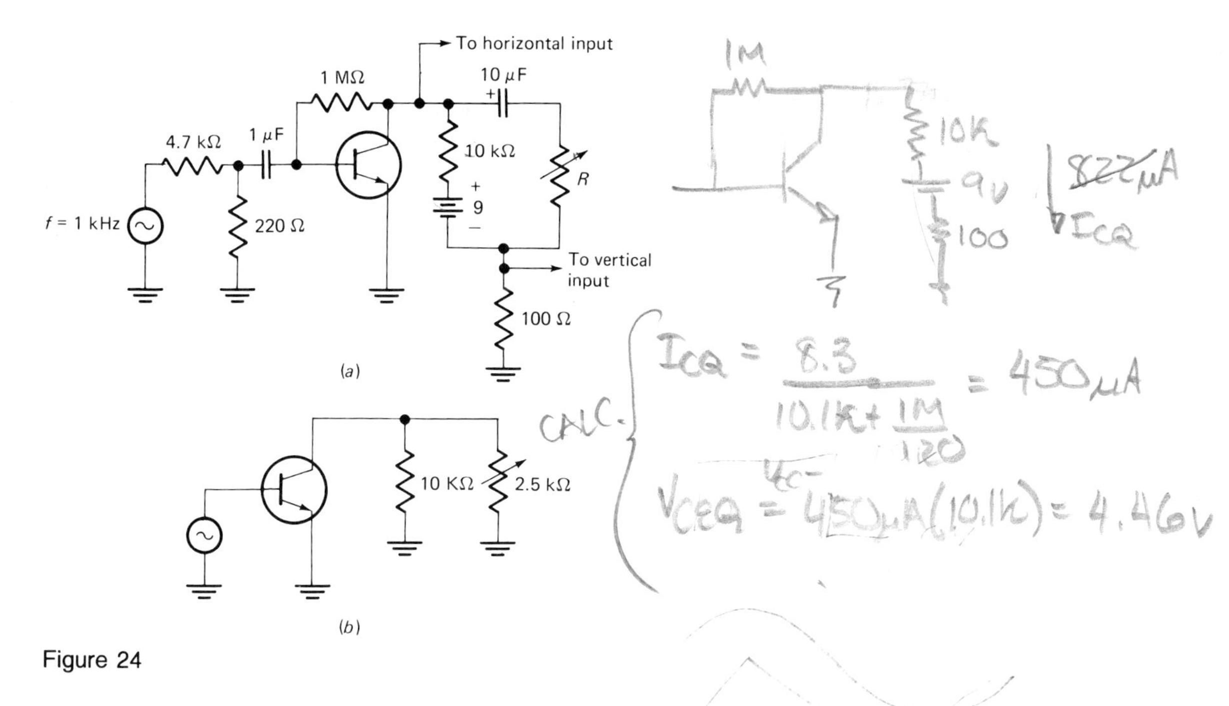

Figure 24

7. For each ac load resistance of Table 2, calculate $I_{C(\text{sat})}$ and $V_{CE(\text{cutoff})}$ using Eqs. (10–4) and (10–5). Record these values in Table 3.

DATA

Table 1. Quiescent Values

I_{CQ}	.4mA / 426μA
V_{CEQ}	4.6V

Table 2. Ac Load Line

R	r_C	$I_{C(\text{sat})}$	$V_{CE(\text{cutoff})}$
2.5 kΩ	2k	2.75	5.4
10 kΩ	5k	1.3	6.8
1 MΩ	9.9k	.9	8.7

Table 3. Calculations

Ac load	$I_{C(\text{sat})}$	$V_{CE(\text{cutoff})}$
1		
2		
3		

25 class a operation

REQUIRED READING

Pages 249–264

EQUIPMENT

1 audio generator
1 power supply: 15 V
1 transistor: 2N3904 (or almost any small-signal *npn* silicon transistor)
5 ½-W resistors: 100 Ω, two 470 Ω, 1 kΩ, 2.2 kΩ
2 capacitors: 10 μF, 100 μF (20-V rating or better)
1 oscilloscope
1 VOM

PROCEDURE

1. In Fig. 25, use a V_{BE} of 0.6 V and round off the 470-Ω resistance to 500 Ω. Calculate I_{CQ} and V_{CEQ}. Record these values in Table 1.
2. With Eq. (10–9), calculate the value of r_E that centers the Q point. Record this value in Table 1.

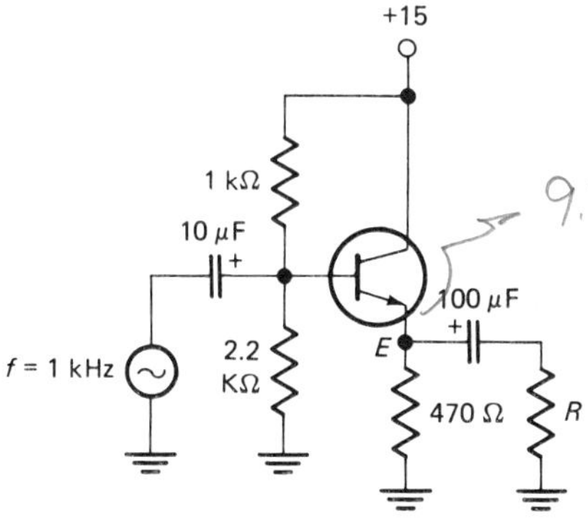

Figure 25

3. Use Eq. (10–11) to work out the quiescent power dissipation of the transistor. Record P_{DQ} in Table 1.
4. Connect the circuit of Fig. 25, using an R of 100 Ω. Adjust the input frequency to 1 kHz.
5. Connect the oscilloscope (dc input) between E and ground. Increase the signal level until clipping just occurs. This corresponds to cutoff clipping (Fig. 10–8*a* in the textbook). If this is what you are getting, write the word "cutoff" in Table 2.
6. Change R to 470 Ω. This results in an r_E that approximately centers the Q point. When you increase the signal level, clipping occurs at both extremes. If this happens, record "both" in Table 2.
7. Change R to infinity (open circuit). When you increase the signal level, you should first get saturation clipping. If so, write "saturation" in Table 2.
8. Change R back to 470 Ω. Increase the signal level until you get the largest possible unclipped signal. Record the peak-to-peak voltage in Table 3.
9. Connect the VOM as an ac voltmeter across R. Measure the rms voltage across R and record the value in Table 3.
10. Calculate the total load power using

$$p_{\text{load}} = \frac{V_{rms}^2}{r_E}$$

Record this value in Table 3.

11. Work out the ratio of p_{load}/P_{DQ} and record in Table 3.

DATA

Table 1. Quiescent Calculations

I_{CQ}	19.4 mA
V_{CEQ}	5.3 V
$r_{E(\text{optimum})}$	273.2
P_{DQ}	103 mW

Table 2. Clipping Effects

R	**Type of clipping**
100 Ω	CUTOFF
470 Ω	BOTH
∞	SATURATION

Table 3. Maximum Signal Swing

V_{pp}	10V
V_{rms}	3.38V
p_{load}	41.8mW
p_{load}/P_{DQ}	.406

26 class b push-pull emitter follower

THEORY

For a current mirror to work properly, the curve of the compensating diode has to match the transconductance curve of the transistor. This is difficult with discrete circuits, and for this reason we will use voltage-divider bias in this experiment. Be especially careful with step 7, because increasing V_{CC} too fast produces excessive collector current, which can destroy either or both transistors.

REQUIRED READING

Pages 272–288

EQUIPMENT

1 audio generator
1 power supply: adjustable from at least 1 to 15 V
1 complementary pair of transistors: 2N3904 and 2N3906 (or a similar complementary pair)
5 ½-W resistors: 100 Ω, two 680 Ω, two 4.7 kΩ
2 capacitors: 1 μF, 100 μF (15-V rating or better)
1 oscilloscope
1 VOM

PROCEDURE

1. Calculate $I_{C(sat)}$ for Fig. 26*a* with a V_{CC} of 10 V. Record the value in Table 1.
2. Calculate 2 percent of $I_{C(sat)}$ and record in Table 1.
3. Connect the circuit of Fig. 26*a* with V_{CC} equal to 5 V.
4. Set the input frequency to 1 kHz and the signal level across the audio generator to 2 V pp.
5. Look at the output signal across the 100-Ω resistor. What kind of distortion is this?
   Write the name here: ______________________________ .

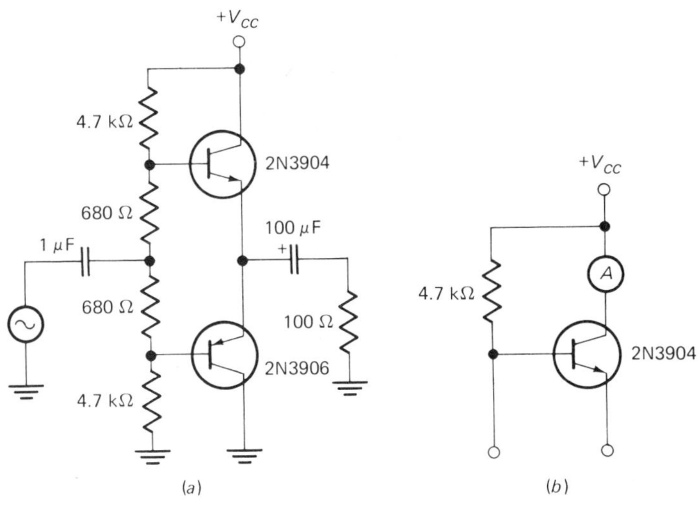

Figure 26

6. Turn the signal level down to zero and connect the VOM as an ammeter in series with the upper collector (see Fig. 26*b*).
7. Slowly increase V_{CC} until I_{CQ} is 1 mA. Remove the VOM and reconnect the upper collector to the V_{CC} supply.
8. Use the VOM to measure V_{BE} (either transistor) and record the value in Table 2.
9. Increase the signal level until you get an output signal of 8 V pp.
10. Slowly increase the signal level until clipping just starts on the output signal.
11. Record the peak-to-peak output voltage in Table 2.
12. Using the VOM as an ac voltmeter, measure the rms value of output voltage. Record in Table 2. Now, calculate and record the value of load power.

DATA

Table 1. $V_{cc} = 10$ V

$I_{C(\text{sat})}$	
2% $I_{C(\text{sat})}$	

Table 2. Measurements

V_{BE}	
V_{pp}	
V_{rms}	
P_{load}	

27 *class c operation*

REQUIRED READING

Pages 298–305

EQUIPMENT

1 audio generator
1 power supply: 10 V
1 transistor: 2N3904 (or almost any small-signal *npn* silicon transistor)
5 ½-W resistors: 10 Ω, 100 Ω, 220 Ω, 2.2 kΩ, 100 kΩ
1 capacitor: 1 μF
1 oscilloscope

PROCEDURE

1. Connect the circuit of Fig. 27*a*.
2. Set the input frequency to 1 kHz. Adjust the signal level to get narrow output pulses with a peak-to-peak value of 6 V. (The output pulses should look like Fig. 12–1*b* in the textbook.)
3. Measure the pulse width W and the period T. Record these values in Table 1.
4. Calculate and record the duty cycle in Table 1.
5. Look at the signal on the base. You should see a negatively clamped sine wave. With the oscilloscope on dc input, measure the positive and negative peaks of this clamped signal. Record both voltages in Table 1.
6. Disconnect the oscilloscope. Set up the controls as follows:
a. Horizontal sensitivity to 1 V/cm (dc input).
b. Vertical sensitivity to 0.1 V/cm (dc input).
c. Center spot in the upper lefthand corner.
7. Connect the circuit of Fig. 27*b* with the frequency at 100 Hz.
8. Increase the signal level until the load line hits the saturation point.
9. Measure $I_{C(\text{sat})}$ and $V_{CE(\text{cutoff})}$ on this load line. Record both values in Table 2.

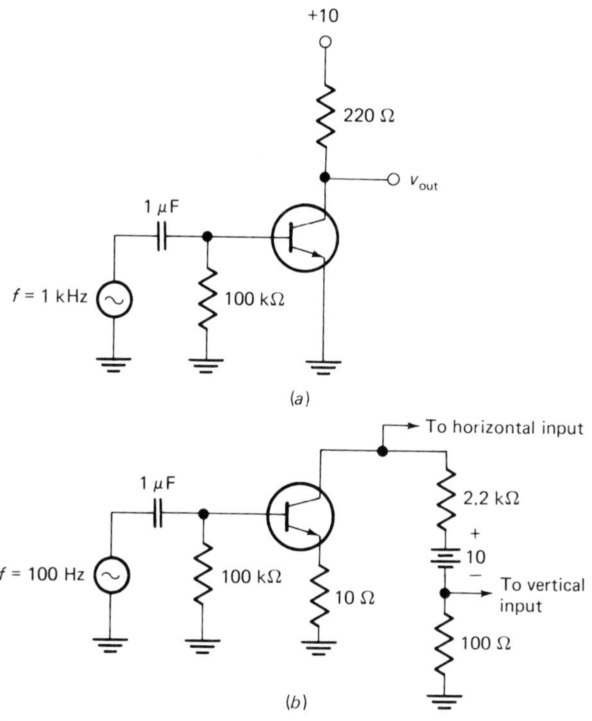

Figure 27

DATA

Table 1. Time Waveforms

W	
T	
duty cycle	
+ peak	
− peak	

Table 2. Load Line

$I_{C(\text{sat})}$	
$V_{CE(\text{cutoff})}$	

28 *JFET curves*

REQUIRED READING

Pages 320–331

EQUIPMENT

1 audio generator
1 power supply: adjustable from at least 1 to 10 V
1 JFET: MPF102 (or any *n*-channel JFET with an I_{DSS} greater than 2 mA)
1 diode: 1N914 (or any small-signal silicon diode)
2 ½-W resistors: 100 Ω, 1 MΩ
1 capacitor: 1 μF
1 oscilloscope

PROCEDURE

1. Set up the oscilloscope as follows:
   a. Horizontal sensitivity to 1 V/cm (dc input).
   b. Vertical sensitivity to 0.1 V/cm (dc input).
   c. Center the spot in the upper lefthand corner.
2. Connect the circuit of Fig. 28*a*.
3. Slowly vary the supply from minimum up to 10 V. If the spot deflects off the screen, change the vertical sensitivity as needed to keep it on.
4. Vary the supply voltage back and forth rapidly from minimum to 10 V. As you do this, notice you get a drain curve similar to Fig. 28*b*.
5. Estimate the pinchoff voltage. Write the value here: $V_P =$ ______________.
6. Adjust the supply to get a V_{DS} greater than pinchoff. Then, figure out the relation between I_{DSS} and vertical deflection. Write the value of I_{DSS} here:

   ______________________.
7. Disconnect the oscilloscope. Center the spot in the upper righthand corner.
8. Connect the circuit of Fig. 28*d*.

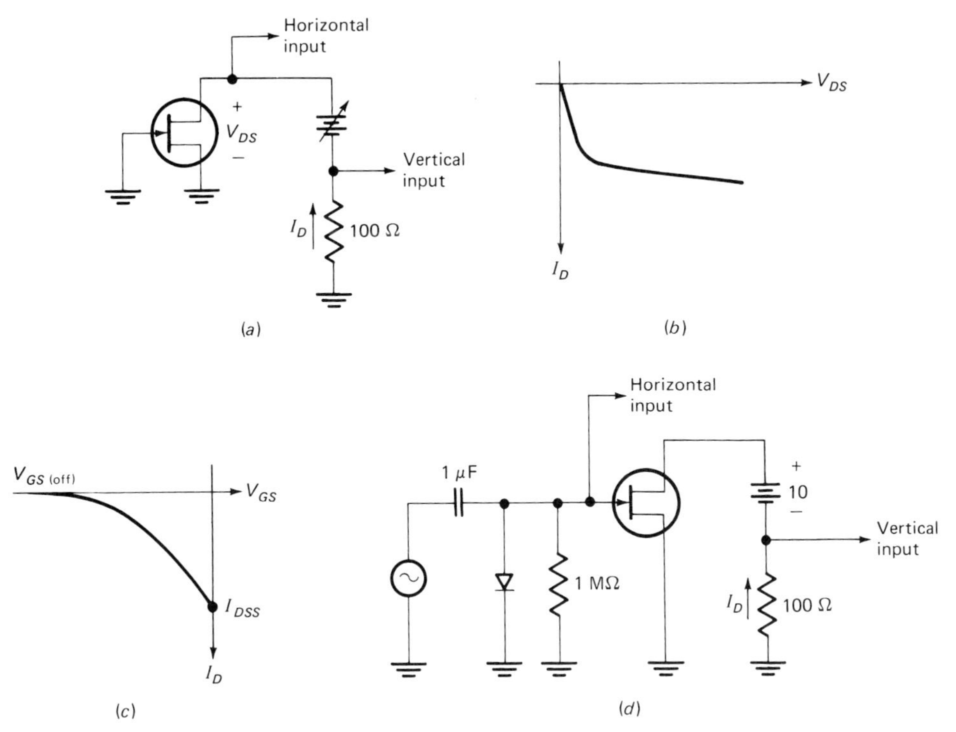

Figure 28

9. Set the audio generator to 100 Hz. Increase the signal from zero upward to get a transconductance curve like Fig. 28*c*. (You may have to readjust sensitivities.)

10. Estimate $V_{GS(\text{off})}$ and record here: ______________________ .

11. Change the horizontal and vertical sensitivity as needed to get as large as a transconductance curve as possible on the screen. Notice how the curve has a parabolic shape.

12. If a curve tracer is available, look at the drain curves of the MPF102.

29 JFET biasing

THEORY

Your textbook covers self-bias and current-source bias, the two most widely used JFET biasing methods. This experiment also introduces two lesser biasing schemes.

Figure 29*a* is called *voltage-divider bias* because a voltage divider supplies the gate. The current through R_S produces a source-to-ground voltage that is slightly greater than the gate-to-ground voltage. For this reason, the gate-source diode is reverse-biased.

Figure 29*c* is called *source bias,* which can be thought of like the emitter bias used with bipolar transistors. The source voltage and source resistor set up a stable value of drain current.

REQUIRED READING

Pages 346–354

EQUIPMENT

2 power supplies: 9 V and 25 V
1 JFET: MPF102 (or any *n*-channel JFET with an I_{DSS} greater than 2 mA)
1 transistor: 2N3904
5 ½-W resistors: one 2.2 kΩ, two 4.7 kΩ, one 220 kΩ, one 470 kΩ
1 VOM

PROCEDURE

1. Connect the voltage-divider bias circuit of Fig. 29*a*. Measure the dc voltage between the gate and ground. Record the value here:

   $V_G =$ ______________________________ .
2. Measure V_S and V_D. Record the values in Table 1.

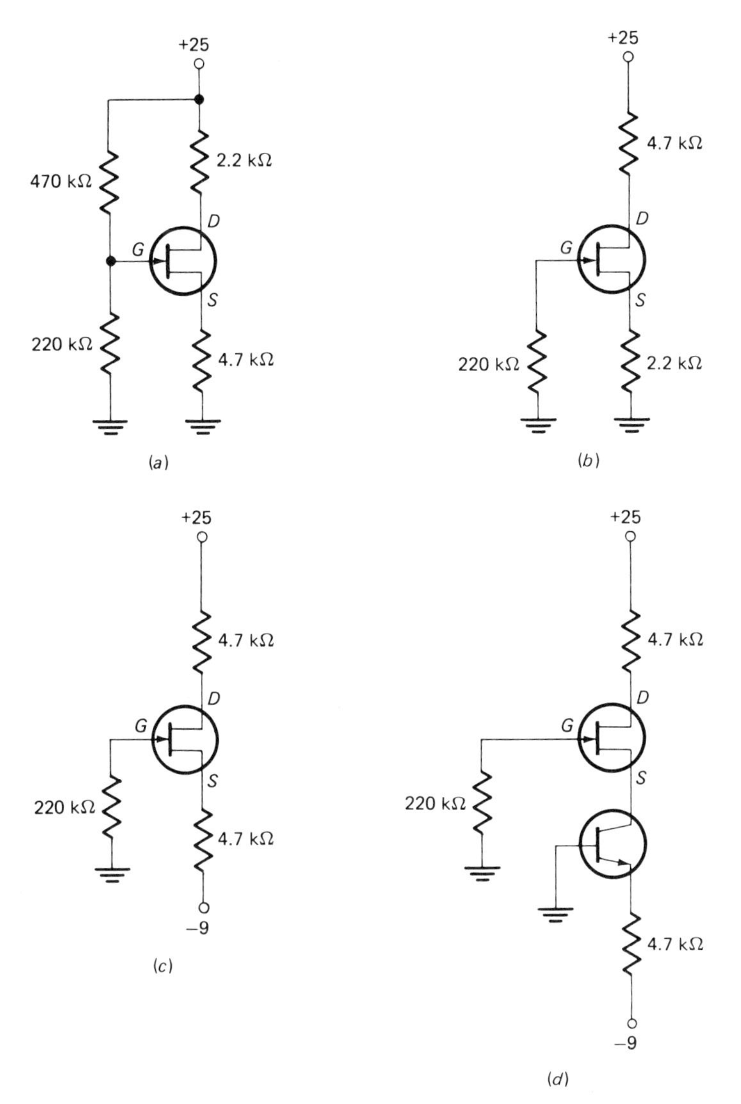

Figure 29

3. Connect the self-bias circuit of Fig. 29*b*. Measure and record the values of V_S and V_D (Table 1).
4. Measure the voltage from gate to ground. Record the value here:

   $V_G =$ ________________________. This should convince you that I_G is negligibly small.
5. Connect the remaining two biasing circuits of Fig. 29. Measure and record the values of V_S and V_D for each (Table 1).

Table 1. Measurements

	Voltage-divider bias	Self-bias	Source bias	Current-source bias
V_S				
V_D				

30 JFET amplifier and source follower

REQUIRED READING

Pages 358–364

EQUIPMENT

1 audio generator
1 power supply: 25 V
1 JFET: MPF 102 (or any *n*-channel JFET with an I_{DSS} greater than 2 mA)
4 ½-W resistors: 1 kΩ, 2.2 kΩ, 4.7 kΩ, 220 kΩ
1 capacitor: 100 μF (10-V rating or better)
1 oscilloscope

PROCEDURE

1. The transconductance of the JFET depends on the type number and quiescent drain current. Using the low and high values given in Table 1, calculate the voltage gain from input to output for the amplifier of Fig. 30*a*. Record this as A_1 in Table 1.
2. Work out the voltage gain for the source follower of Fig. 30*b* and record the values as A_2 in Table 1.
3. Connect the JFET amplifier of Fig. 30*a*.
4. Adjust the audio generator to 1 kHz. Set the signal level at 0.1 V pp across the 220-kΩ resistor.
5. Look at the output signal. It should be an amplified sine wave. Record the peak-to-peak value of this signal in Table 2.
6. Calculate and record the voltage gain in Table 2.
7. Use the A of Table 2 and Eq. (14–9) to find the value of g_m. Record this value in Table 2.
8. With the oscilloscope still across the output, open the bypass capacitor. Record the new peak-to-peak value of output voltage in Table 2.
9. Connect the source follower of Fig. 30*b*.
10. Adjust the frequency to 1 kHz and set the signal level to 1 V pp across the 220-kΩ resistor.

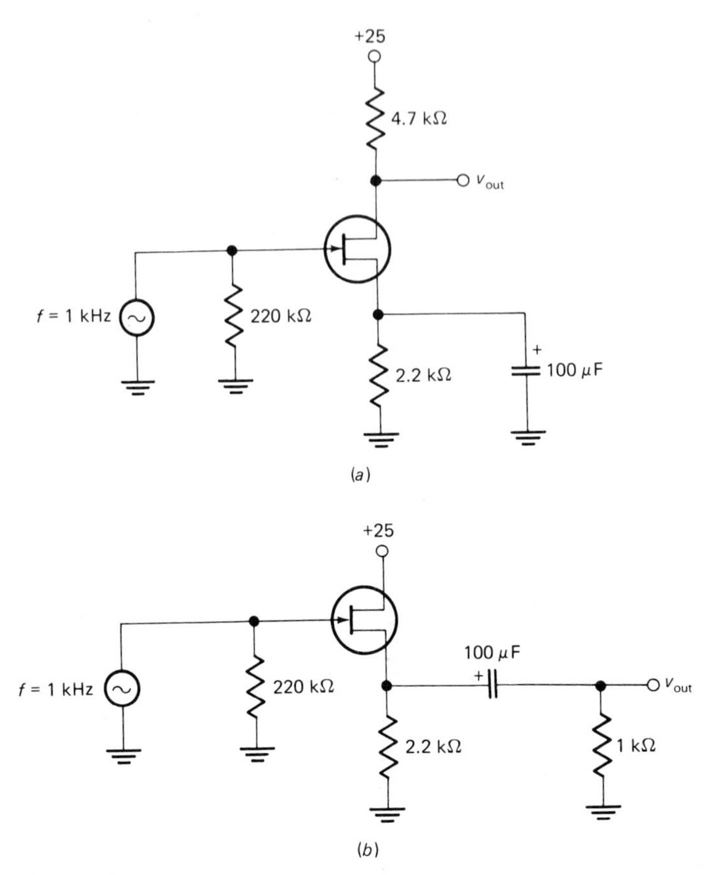

Figure 30

11. Look at the output signal of the source follower. Record the peak-to-peak voltage in Table 3.
12. Calculate the voltage gain A and record this value in Table 3.

DATA

Table 1. Calculations

g_m	A_1	A_2
1500 μS		
6000 μS		

Table 2. JFET Amplifier

$v_{in} = 0.1$ V pp	
v_{out}	______
A	______
g_m	______
v_{out} (no bypass)	______

Table 3. Source Follower

$v_{in} = 1$ V pp	
v_{out}	______
A	______

31 decibels

REQUIRED READING

Pages 379–393

EQUIPMENT

1 audio generator
1 decade resistance box
7 ½-W resistors: four 470 Ω, two 1 kΩ, 100 kΩ
1 ac millivoltmeter

PROCEDURE

1. Figure 31*a* shows a voltage divider. For each value of R given in Table 1, calculate the bel voltage gain. Round off your answers to the nearest decibel and record the values in Table 1.
2. Connect the circuit of Fig. 31*a* using an R of 240 kΩ.

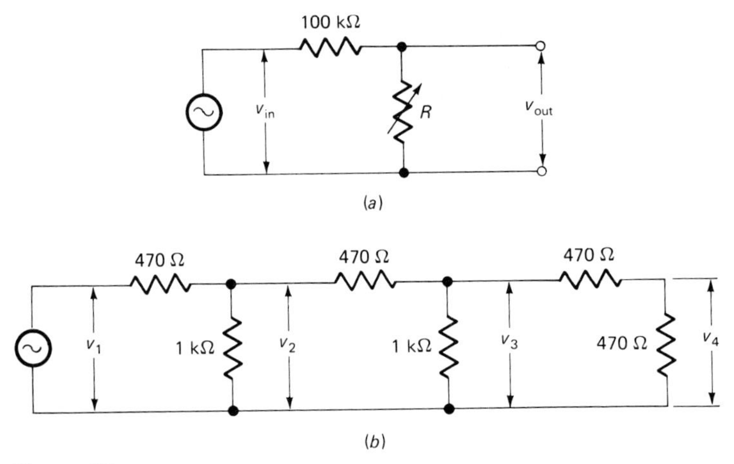

Figure 31

3. Put the ac millivoltmeter on the 1-V range. Adjust the audio generator to get a v_{in} reading of 0 dB.
4. Measure v_{out} using the dB scale. This reading is the bel voltage gain. Record the value in Table 1 (round off to nearest decibel).
5. Repeat steps 3 and 4 for the other R values listed in Table 1.
6. Figure 31b shows three cascaded voltage dividers. Let $A_1 = v_2/v_1$, $A_2 = v_3/v_2$, and $A_3 = v_4/v_3$. Round off 470 to 500 and calculate the bel voltage gain A_1', A_2', and A_3'. Record in Table 2.
7. Add these bel voltage gains to get the total bel voltage gain. Record this value as A' in Table 2.
8. Connect the circuit of Fig. 31b.
9. Set v_1 at 0 dB on the 1-V range of the ac millivoltmeter.
10. Read the values of v_2, v_3, and v_4 rounded off to the nearest decibel and record in Table 3.

DATA

Table 1.

R	Calculated A'	Measured A'
240 kΩ		
100 kΩ		
46 kΩ		
11.1 kΩ		
1 kΩ		

Table 2.

A_1'	______
A_2'	______
A_3'	______
A'	______

Table 3.

$v_1 = 0$ dB	
v_2	______
v_3	______
v_4	______

32 *Miller's theorem*

REQUIRED READING

Pages 393–398

EQUIPMENT

1 audio generator
2 transistor: 2N3904 (or almost any small-signal *npn* silicon transistor)
8 ½-W resistors: 10 Ω, two 100 Ω, 1 kΩ, 4.7 kΩ, 10 kΩ, 100 kΩ, 10 MΩ
4 capacitors: 0.001 μF, 0.01 μF, two 1 μF
1 oscilloscope

PROCEDURE

1. In the Darlington amplifier of Fig. 32*a*, the voltage gain from base to collector is

$$A \cong \frac{1000}{110} \cong 9$$

   Use this value of A in Fig. 32*b* to calculate $R_{\text{in(Miller)}}$. Record the result in Table 1.
2. Similarly, use an A of 9 in Fig. 32*c* to calculate $C_{\text{in(Miller)}}$. Record the answer in Table 1.
3. Connect the Darlington amplifier of Fig. 32*a*. This basic amplifier will be used for the remainder of the experiment.
4. Connect the circuit of Fig. 32*b* with the Darlington amplifier inside the imaginary box.
5. Adjust the audio generator to 1 kHz. Set the signal level to get an amplified output of 1 V pp
6. Look at v_{in} and record its peak-to-peak value in Table 2.
7. Look at v_A and record its peak-to-peak value in Table 2.
8. Connect the circuit of Fig. 32*c* with the Darlington amplifier inside the box.
9. Repeat steps 5 through 7, recording your results in Table 3.

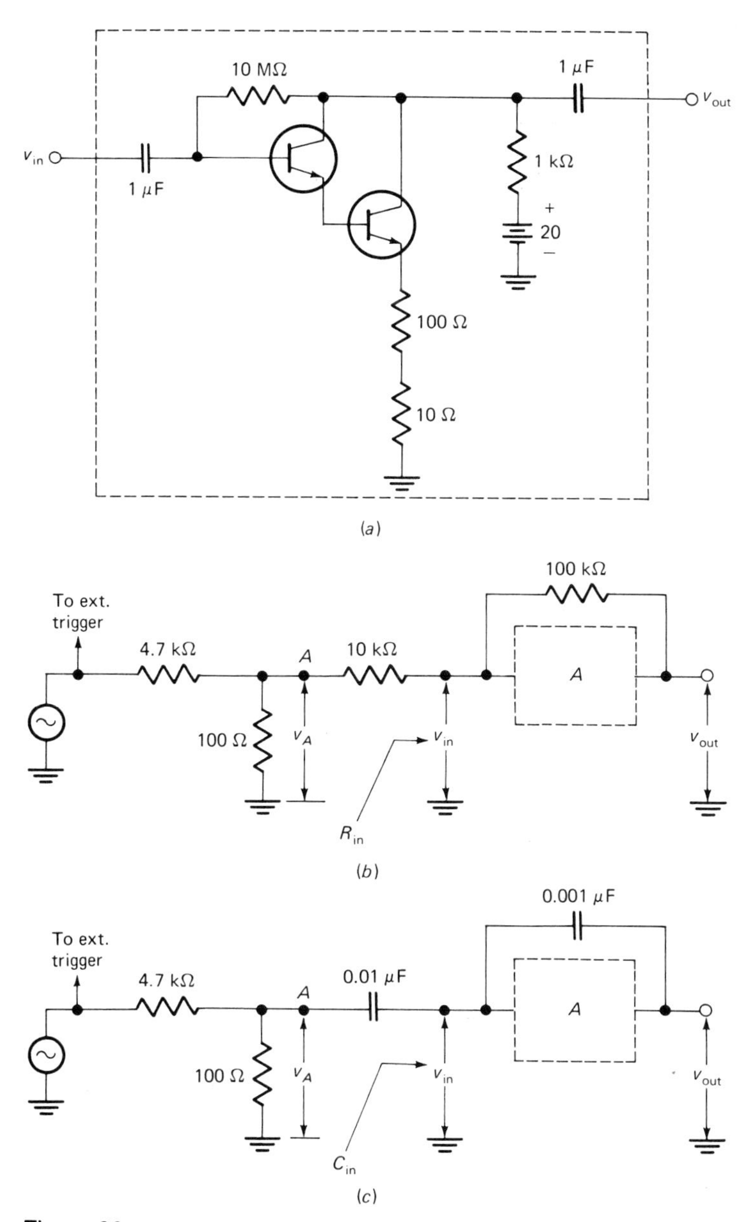
10 MΩ
1 μF
v_{out}
v_{in}
1 μF
1 kΩ
+
20
−
100 Ω
10 Ω
(a)
100 kΩ
To ext.
trigger
4.7 kΩ
A
10 kΩ
A
100 Ω
v_A
v_{in}
v_{out}
R_{in}
(b)
0.001 μF
To ext.
trigger
4.7 kΩ
A
0.01 μF
A
100 Ω
v_A
v_{in}
v_{out}
C_{in}
(c)

Figure 32

DATA

Table 1. Calculations for $A = 9$

$R_{in(Miller)}$	
$C_{in(Miller)}$	

Table 2. Resistance

$v_{out} = 1$ V pp	
v_{in}	
v_A	

Table 3. Capacitance

$v_{out} = 1$ V pp	
v_{in}	
v_A	

33 *the lag network*

REQUIRED READING

Pages 412–430

EQUIPMENT

1 audio generator
1 ½-W resistor: 10 kΩ
1 capacitor: 0.01 μF
1 ac millivoltmeter
1 oscilloscope

PROCEDURE

1. Calculate the critical frequency of the lag network shown in Fig. 33*a*. Record the value here:

   $f_c =$ ____________________

2. Connect the circuit of Fig. 33*a*. Set the frequency at the f_c of the preceding step. Connect the ac millivoltmeter across the input and set the signal level to read 0 dB on the 1-V range.
3. Measure the output voltage with the ac millivoltmeter. If the R and C of the lag network have exact values, the output will be 3 dB lower than the input. If necessary, readjust the frequency slightly to get a 3-dB difference between v_{in} and v_{out}. Record the final frequency at the top of Table 1.
4. Set the frequency to $0.1f_c$ (use the f_c at the top of Table 1).
5. Read the input and output voltage in decibels. Record these values in Table 1. Calculate and record the bel voltage gain A'.
6. Repeat step 5 for each frequency in Table 1.
7. Connect the oscilloscope to the lag network as shown in Fig. 33*b*. Set the frequency to $0.1f_c$.
8. Measure the period T. Record this value in Table 2. (You may need to increase signal level to get external triggering.)

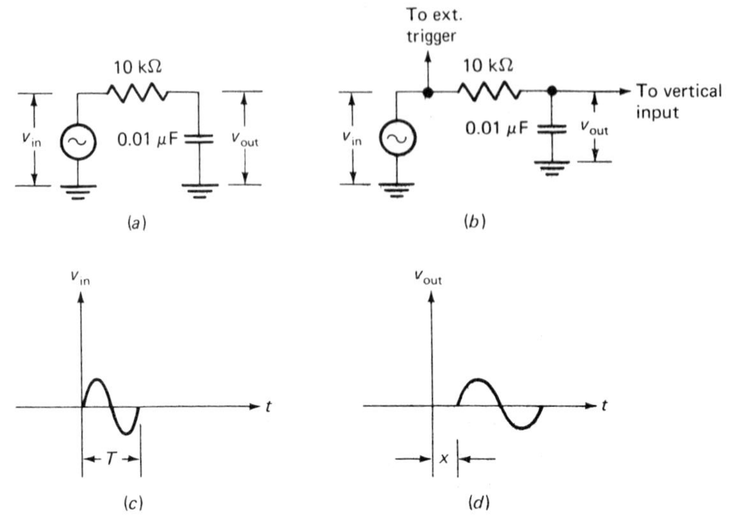

Figure 33

9. Look at the input signal and adjust the centering controls of the oscilloscope until the start of the sine wave passes through the center of the screen (see Fig. 33*c*).
10. Look at the output signal and measure lag x (see Fig. 33*d*). If necessary, increase vertical sensitivity. Record x in Table 2.
11. Calculate and record the phase angle using

$$\phi = -\frac{x}{T} 360^\circ$$

12. Repeat steps 8 through 11 for the other frequencies in Table 2.

DATA

Table 1. $f_c =$ ______________

f	V_{in}	V_{out}	A'
$0.1f_c$			
f_c			
$10f_c$			
$100f_c$			

Table 2.

f	T	x	ϕ
$0.1f_c$			
f_c			
$10f_c$			

34 JFET high-frequency analysis

REQUIRED READING

Pages 430–437

EQUIPMENT

1 audio generator
1 power supply: 25 V
1 JFET: MPF102 (or any *n*-channel JFET with an I_{DSS} greater than 2 mA)
4 ½-W resistors: 100 Ω, 2.2 kΩ, 4.7 kΩ, 10 kΩ
4 capacitors: three 1000 pF, 100 μF (10-V rating or better)
1 ac millivoltmeter
1 oscilloscope

PROCEDURE

1. To simplify this experiment we add large capacitors to the JFET as shown in Fig. 34. This brings the upper cutoff frequency down low enough to measure easily. Because internal FET capacitances are normally less than 10 pF, the effective values to use in this experiment become

$$C_{gs} = C_{gd} = C_{ds} \cong 1000 \text{ pF}$$

2. In Fig. 34, work out the approximate value of R_S in the gate lag network. Using a g_m of 2000 μS, calculate the effective C for the gate lag network. Record R_S and C in Table 1.
3. Work out the critical frequency of the gate lag network and record the value in Table 1.
4. In Fig. 34, what is the ac drain resistance r_D? Record the answer in Table 2.
5. Calculate the effective capacitance and the critical frequency of the drain lag network. Record the values in Table 2.
6. Connect the JFET amplifier of Fig. 34.
7. Set the frequency to 100 Hz and the signal level to get a v_{out} of 1 V pp on the oscilloscope. The waveform should be a sine wave. If so, disconnect the oscilloscope.

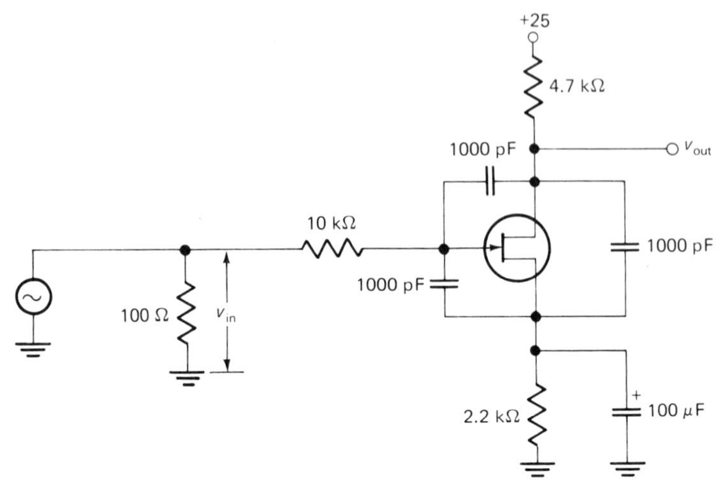

Figure 34

8. Connect the ac millivoltmeter across the output. With the frequency at 100 Hz, set the signal level to get a v_{out} of 1 V rms.
9. Measure the rms value of v_{in} (Fig. 34). Calculate the voltage gain A. Record these values in Table 3.
10. Again measure the input and output voltage but this time read voltages in decibels. Record the bel voltage gain A' in Table 3.
11. The gate lag network is dominant (see Tables 1 and 2). The critical frequency of this network should be in the vicinity of the f_c in Table 1. Find the actual critical frequency by locating the frequency where the bel voltage gains is down 3 dB from the value in Table 3. Record the actual f_c of the gate lag network in Table 4.
12. Short out the 10-kΩ resistor. This removes the gate lag network. Find the critical frequency of the drain lag network by locating the frequency where the bel voltage gain is down 3 dB from the value of Table 3. Record the value of f_c for the drain lag network in Table 4.

DATA

Table 1. Gate Lag Network

R_S	
C	
f_c	

Table 2. Drain Lag Network

r_D	
C	
f_c	

Table 3. $f = 100$ Hz

$v_{out} = 1$ V rms	
v_{in}	
A	
A'	

Table 4. Critical f's

lag network	f_c
gate	
drain	

35 risetime and sagtime

REQUIRED READING

Pages 427–429, 446–447

EQUIPMENT

1 sine/square generator
1 ½-W resistor: 22 kΩ
1 capacitor: 0.01 μF
1 ac millivoltmeter
1 oscilloscope

PROCEDURE

1. Calculate the critical frequency of the lag network shown in Fig. 35*a*. Record this value in Table 1.

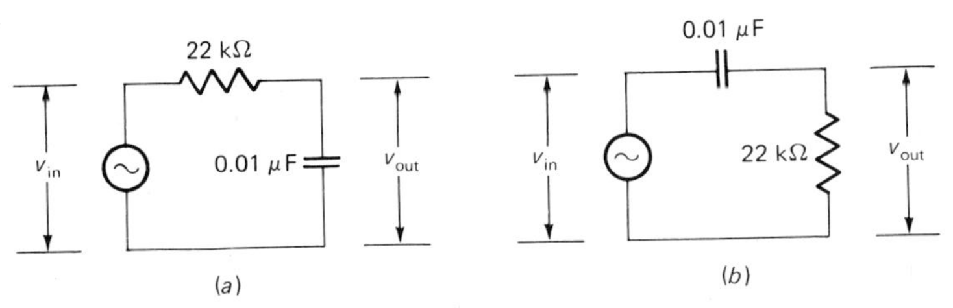

Figure 35

2. Use Eq. (16–13) to get the risetime of the output for a step input in Fig. 35*a*. Record the answer in Table 1.
3. Connect the circuit of Fig. 35*a*. Adjust the generator to deliver a sine wave. Set the frequency to the f_c in Table 1 and the signal level to a v_{in} of 0 dB on the 1-V range of the ac millivoltmeter.
4. Read the output voltage in decibels. Adjust the frequency slightly until it is down 3 dB from the input. Record the measured critical frequency in Table 2.

5. Put a 300-Hz square wave into the lag network. Look at the input signal with the oscilloscope and set the level at 1 V pp.
6. Measure the risetime of the output signal. Record T_R in Table 2.
7. Calculate the critical frequency of the lead network (Fig. 35*b*). Record in Table 3.
8. Use Eq. (16–26*b*) to get the sagtime of the output. Record T_S in Table 3.
9. Connect the circuit of Fig. 35*b*. Set the frequency to the f_c in Table 3. Put a sine wave into the lead network and adjust the level of v_{in} at 0 dB on the 1-V range of the ac millivoltmeter.
10. Read the output voltage in decibels. Adjust the frequency slightly until the output is down 3 dB from the input. Record this break frequency in Table 4.
11. Put a 300-Hz square wave into the lead network. Look at the input signal with the oscilloscope and set the signal level at 1 V pp.
12. Measure the sagtime of the output signal and record T_S in Table 4.

DATA

Table 1. Lag Calculations

$f_c =$	
$T_R =$	

Table 2. Lag Measurements

$f_c =$	
$T_R =$	

Table 3. Lead Calculations

$f_c =$	
$T_S =$	

Table 4. Lead Measurements

$f_c =$	
$T_S =$	

36 the differential amplifier

REQUIRED READING

Pages 461–476

EQUIPMENT

1 audio generator
2 power supplies: 9 V and 25 V
2 transistors: 2N3904 (if necessary, you can substitute almost any small-signal *npn* silicon transistor)
6 ½-W resistors: two 100 Ω, 4.7 kΩ, two 10 kΩ, 100 kΩ
1 ac millivoltmeter
1 VOM
1 oscilloscope

PROCEDURE

1. In Fig. 36*a*, work out the value of r_e'. Record in Table 1.
2. Calculate the theoretical voltage gain using R_C/r_e'. Record the value in Table 1.
3. Connect the circuit of Fig. 36*a*.
4. Use the oscilloscope to look at v_{c2} (the ac collector-to-ground voltage). Adjust the audio generator to 1 kHz and the signal level to get a v_{c2} and 2 V pp.
5. Because the external trigger is taken from the audio generator, you can look at the phases of signals in the diff amp. Look at each signal listed in Table 2 and record its phase with respect to v_1.
6. Use the ac millivoltmeter to measure the rms value of each voltage in Table 3. (All voltages are measured with respect to ground.) Record all values in Table 3.
7. Table 2 tells you v_{c1} and v_{c2} are 180° out of phase. Table 3 tells you they are approximately equal in amplitude. Because of this, the output voltage is a sine wave of the same frequency but approximately twice the amplitude of v_{c1} or v_{c2} (see Fig. 36*b*). Based on this information, write the rms value of v_{out} in Table 4.

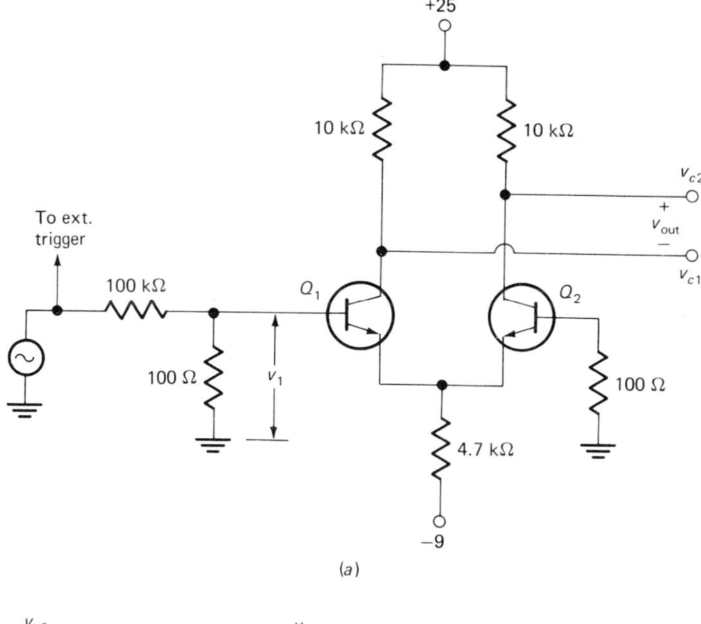

(a)

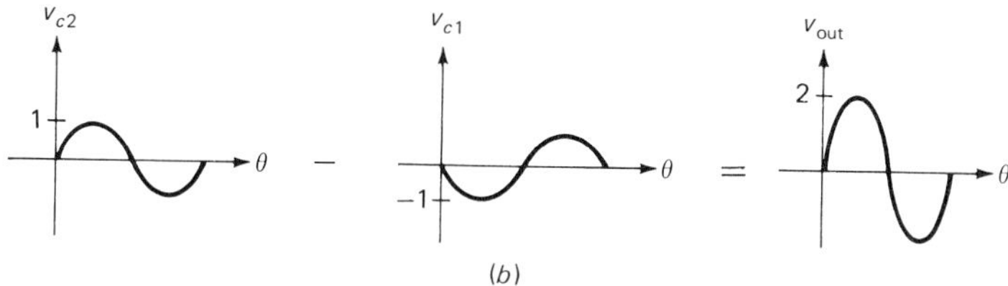

(b)

Figure 36

8. Transcribe the v_1 value of Table 3 to Table 4. Calculate and record the voltage gain A.
9. Use the VOM to measure the dc voltage between collectors. Record this as $V_{\text{out(offset)}}$ in Table 5. Calculate the input offset voltage using the A of Table 4 and

$$V_{\text{in(offset)}} = \frac{V_{\text{out(offset)}}}{A}$$

Record your answer in Table 5.

DATA

Table 1.

r'_e	
A	

Table 2.

v_{c2} is	
v_{c1} is	
v_1 is in phase	
v_e is	

Table 3.

v_{c2}	
v_{c1}	
v_1	
v_e	

Table 4.

v_{out}	
v_1	
A	

Table 5.

$V_{out(offset)}$	
$V_{in(offset)}$	

37 the operational amplifier

REQUIRED READING

Pages 476–486

EQUIPMENT

2 power supplies: 15 V
2 op amps: 741C, 318C (DIL-8 or TO-5 packages)
4 ½-W resistors: two 1 kΩ, 10 kΩ, 100 kΩ
2 capacitors: 1 μF (25-V rating or better)
1 sine/square generator
1 oscilloscope
1 ac millivoltmeter

PROCEDURE

1. Build the circuit of Fig. 37 using a 741C and an R of 10 kΩ. Connect the 1-μF bypass capacitors as close to the IC as possible to prevent oscillations (discussed in Chap. 19).
2. Connect the oscilloscope to display the output of the op amp. Adjust the sine/square generator to get 20-V peak-to-peak sine wave on the oscilloscope at 1 kHz.
3. Measure the input signal (pin 3) with the ac millivoltmeter and record the rms value in Table 1. Measure and record the output signal (pin 6).
4. Change R to 100 kΩ and repeat steps 2 and 3.
5. Drive the input with a square wave instead of a sine wave. Adjust the amplitude until slew-rate limiting occurs (the oscilloscope time base should be in the vicinity of 10 μs/cm). Increase the amplitude and notice that the slew rate remains approximately the same.
6. Record the slew rate in Table 2.
7. Change R to 10 kΩ and adjust the input amplitude to get slew-rate limiting. Notice that the slew rate is still approximately the same as before.
8. Change from square-wave input to sine-wave input. Adjust the signal level to get 20 V peak-to-peak on the oscilloscope.

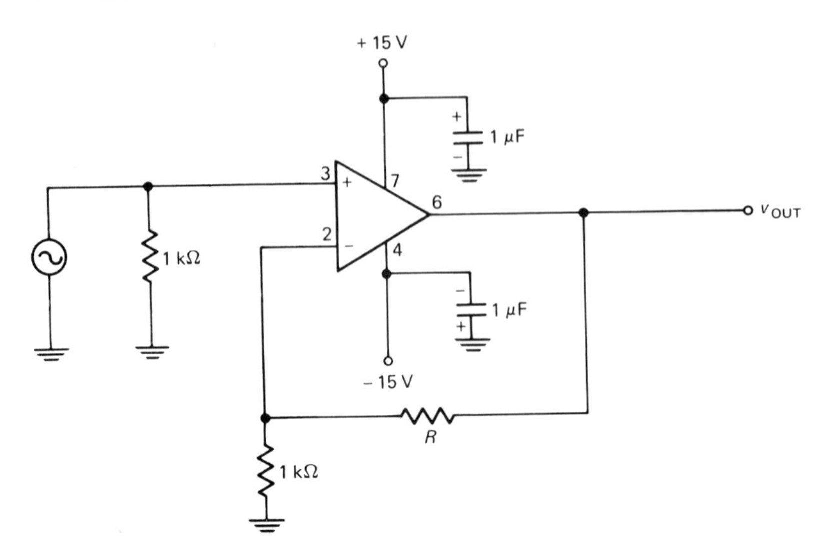

Figure 37

9. Increase the frequency from 1 kHz upward until slew-rate distortion occurs. Record the approximate frequency where this distortion begins (use Table 2, f_{max}).
10. Change from a 741C to a 318C and repeat steps 2 through 9, recording the results in Tables 3 and 4.

DATA

Table 1. 741C Gain

R	10 kΩ	100 kΩ
V_{IN}		
V_{OUT}		

Table 2. 741 Slew Rate and Power Bandwidth

S_R	
f_{max}	

Table 3. 318C Gain

R	10 kΩ	100 kΩ
V_{IN}		
V_{OUT}		

Table 4. 318C Slew Rate and Power Bandwidth

S_R	
f_{max}	

38 sp negative feedback

REQUIRED READING

Pages 495–508

EQUIPMENT

1 audio generator
2 power supplies: 9 V
1 op amp: 741C (DIL-8 or TO-5 package)
6 ½-W resistors: three 1 kΩ, two 100 kΩ, 1 MΩ (if possible, use precision resistors)
1 potentiometer: 5 kΩ (or nearest available)
1 ac millivoltmeter
1 VOM
1 oscilloscope

PROCEDURE

1. Calculate the value of A_{SP} in Fig. 38*a*. If v_{in} is 10 mV rms, what is the approximate value of v_f? Of v_{out}? Record the answers in Table 1.
2. Connect the circuit of Fig. 38*a*. Adjust the frequency to 1 kHz and v_{out} to 3 V pp.
3. Look at v_{in} and v_f. Notice they are sine waves like the output except they have peak-to-peak values of approximately 30 mV pp. Also, v_{in}, v_f, and v_{out} are in phase.
4. Connect the ac millivoltmeter across the output. Set the signal level to 1 V rms.
5. Use the ac millivoltmeter to measure v_{in} and v_f. Record these rms values in Table 2.
6. Use the measured v_{out} and v_{in} to calculate the A_{SP} of the amplifier. Record this value in Table 2.
7. Connect the circuit of Fig. 38*b* with the VOM used as a dc voltmeter across the output. (Leave the 1-MΩ resistor open as shown.)
8. Adjust the potentiometer to get zero output voltage.

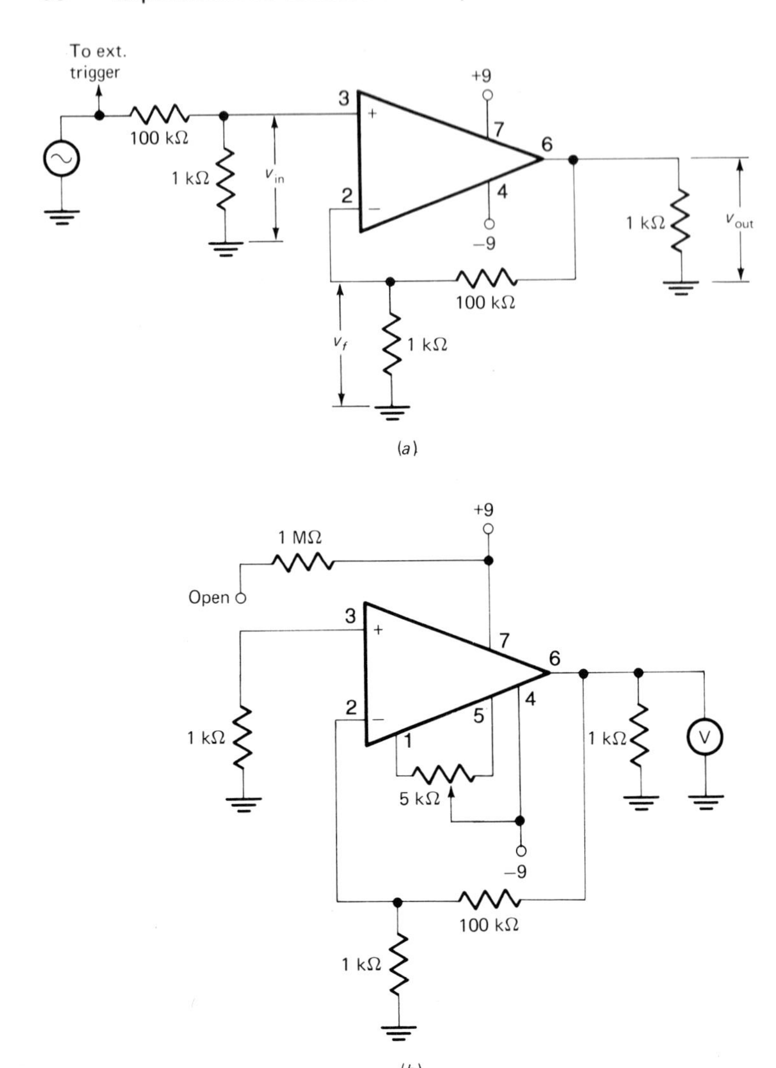

To ext.
trigger
100 kΩ
1 kΩ
V_{in}
3
+
2
−
+9
7
6
4
−9
1 kΩ
V_{out}
100 kΩ
V_f
1 kΩ
(a)
+9
1 MΩ
Open
3
+
7
6
2
−
5
4
1
1 kΩ
1 kΩ
V
5 kΩ
−9
100 kΩ
1 kΩ
(b)

Figure 38

9. Connect the 1-MΩ resistor to the noninverting input. Record the VOM reading here: V_{out} = ______________.

DATA

Table 1.

A_{SP}	______
v_{in} = 10 mV rms	
v_f	______
v_{out}	______

Table 2.

A_{SP}	______
v_{in}	______
v_f	______
v_{out} = 1 V rms	

39 *pp and ss negative feedback*

REQUIRED READING

Pages 516–523

EQUIPMENT

2 power supplies: 9 V
1 op amp: 741C (DIL-8 or TO-5 package)
6 ½-W resistors: 10 Ω, 100 Ω, two 1 kΩ, 100 kΩ, 1 MΩ
1 potentiometer: 5 kΩ (or nearest available)
1 VOM
1 oscilloscope

PROCEDURE

1. In Fig. 39*a*, how much output voltage is there when the input current equals 10 μA? Record the answer here: V_{out} = ______________________.
2. Connect the circuit of Fig. 39*a*. Connect one end of the 100-kΩ resistor to +9 V, but leave the bottom end open as shown.
3. The vertical sensitivity should be set at 100 mV/cm, dc input.
4. Connect the bottom of the 100-kΩ resistor to the inverting input. You should get a negative dc output voltage. Record the value here:

   v_{out} = ______________________.
5. If v_{in} is 1 mV in Fig. 39*b*, how much is i_{out}? Write the answer here:

   i_{out} = ______________________.
6. Connect the circuit of Fig. 39*b* with the VOM used as an ammeter.
7. Adjust the potentiometer to null the output current.
8. Connect the bottom of the 1-MΩ resistor to the noninverting input. Record the VOM reading here: i_{out} = ______________________.

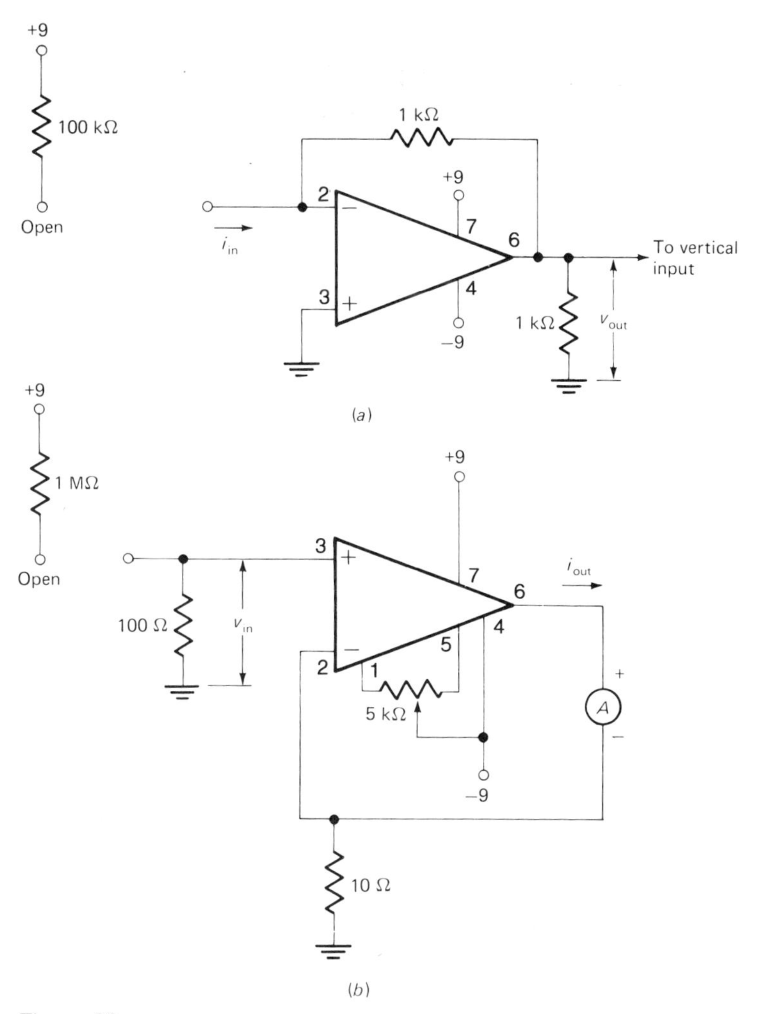
+9
100 kΩ
Open
1 kΩ
+9
2
7
6
To vertical input
3
4
1 kΩ
v_{out}
i_{in}
−9
(a)
+9
1 MΩ
Open
+9
3
7
6
i_{out}
100 Ω
v_{in}
4
5
2
1
5 kΩ
A
−9
10 Ω
(b)

Figure 39

40 *other negative feedback circuits*

REQUIRED READING

Pages 524–527, 591–594

EQUIPMENT

2 audio generators
2 power supplies: 10 V
1 op amp: 741C (DIL-8 or TO-5 package)
8 ½-W resistors: 100 Ω, two 1 kΩ, 2.2 kΩ, 10 kΩ, 100 kΩ, 470 kΩ, 10 MΩ
1 potentiometer: 5 kΩ (or nearest available)
1 VOM
1 oscilloscope

PROCEDURE

1. In Fig. 40*a*, what is the current through the 10-MΩ resistor if the bottom end of this resistor is grounded? Write the answer in Table 1 next to i_1.
2. If the bottom end of the 470-kΩ resistor is grounded, what is the current through this resistor? Write the answer in Table 1 next to i_2.
3. What is the current gain of the PS negative-feedback amplifier shown in Fig. 40*a*? Write the answer in Table 1.
4. Ideally, what does the ammeter of Fig. 40*a* read when the input currents equal the i_1 and i_2 of Table 1? Record your answers in Table 1.
5. Connect the circuit of Fig. 40*a* using the VOM as an ammeter.
6. Null the output current by adjusting the potentiometer. Connect the bottom of the 10-MΩ resistor to the inverting input. Record the ammeter reading in Table 2 next to $i_{out(1)}$.
7. Disconnect the 10-MΩ resistor from the inverting input. Change the ammeter to higher range. Connect the bottom end of the 470-kΩ resistor to the inverting input. Record the ammeter reading in Table 2 next to $i_{out(2)}$.
8. Each channel of Fig. 40*b* acts like a multiplier. The final output is the sum of the amplified inputs. Calculate the voltage gain of the upper channel (v_{out}/v_1). Write the answer here: $A_1 =$ ____________________.

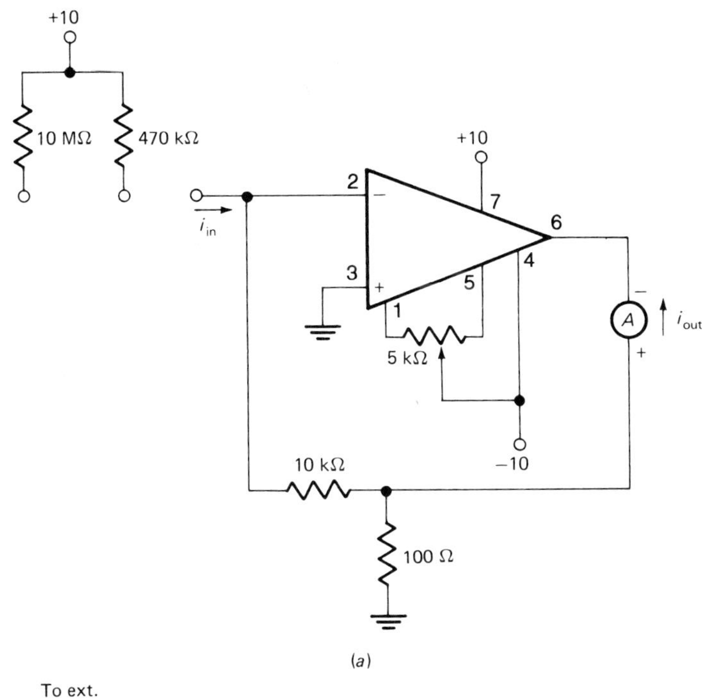

(a)

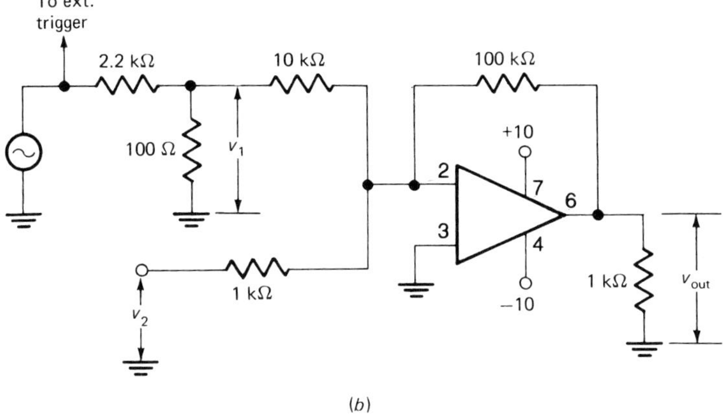

(b)

Figure 40

9. Calculate the voltage gain of the lower channel (v_{out}/v_2) and write the answer here: $A_2 =$ ________________________.

10. Connect the circuit of Fig. 40*b*.

11. Set the audio generator to 1 kHz. Adjust the signal level to get a v_1 of 0.2 V pp on the oscilloscope.

12. Look at the output signal. Notice it is out of phase with the input. Record the peak-to-peak output here: $v_{out(1)} =$ ____________. Calculate the voltage gain of the upper channel and write the answer here: $A_1 =$ ____________.
13. With 1 kHz still driving channel 1, connect the second audio generator to the lower input. Set the frequency to 20 kHz and slowly increase the signal level from zero until the peak-to-peak level is 4 V at the final output.
14. Adjust either frequency until the 20 kHz is almost standing still. This is the kind of signal you get when one sine wave is added to another sine wave.

DATA

Table 1. Calculations

i_1	
i_2	
β_{PS}	
$i_{out(1)}$	
$i_{out(2)}$	

Table 2. Measurements

$i_{out(1)}$	
$i_{out(2)}$	

41 gain-bandwidth product

REQUIRED READING

Pages 508–516

EQUIPMENT

1 square wave generator
2 power supplies: 10 V
1 op amp: 741C (DIL-8 or TO-5 package)
6 ½-W resistors: 1 kΩ, 2.2 kΩ, 10 kΩ, 22 kΩ, 47 kΩ, 100 kΩ
1 oscilloscope

PROCEDURE

1. For each R in Table 1, calculate the A_{SP} in Fig. 41. Record all answers.
2. The typical gain-bandwidth product (same as f_{unity} or $A_{\text{mid}}f_2$) of a 741C is in the vicinity of 1 MHz. Use this value and Eq. (18–16) to work out the $f_{2(SP)}$ for each A_{SP} in Table 1. Record values.
3. Use Eq. (16–13) to find the output risetime for each $f_{2(SP)}$ of Table 1. Record all risetimes.

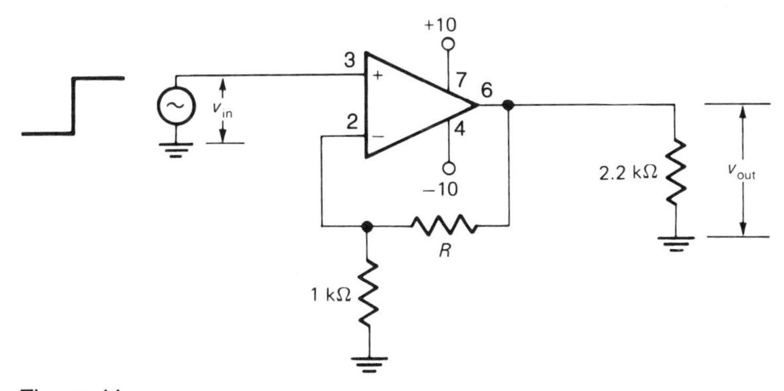

Figure 41

4. Connect the circuit of Fig. 41 using an R of 100 kΩ.
5. With the oscilloscope across the output 2.2-kΩ resistor, set the frequency to 1 kHz and the signal level to 3 V pp.
6. Measure the risetime and record the value in Table 2.
7. Calculate the value of $f_{2(SP)}$ using Eq. (16–13) and the measured risetime. Record this frequency in Table 2.
8. Repeat steps 5 through 7 for each of the remaining R values of Table 2.

DATA

Table 1. Calculations for f_{unity} = 1 MHz

R	A_{SP}	$f_{2(SP)}$	T_R
100 kΩ			
47 kΩ			
22 kΩ			
10 kΩ			

Table 2. Measurements

R	T_R	$f_{2(SP)}$
100 kΩ		
47 kΩ		
22 kΩ		
10 kΩ		

42 *the Wien-bridge oscillator*

REQUIRED READING

Pages 539–543

EQUIPMENT

2 power supplies: 9 V
1 op amp: 741C (DIL-8 or TO-5 package)
7 ½-W resistors: 1 kΩ, two 2.2 kΩ, two 4.7 kΩ, two 10 kΩ
1 potentiometer: 5 kΩ (or nearest available more than 2 kΩ)
2 capacitors: 0.01 μF
1 oscilloscope

PROCEDURE

1. Calculate the frequency of oscillation in Fig. 42 for an R of 10 kΩ. Record in Table 1.

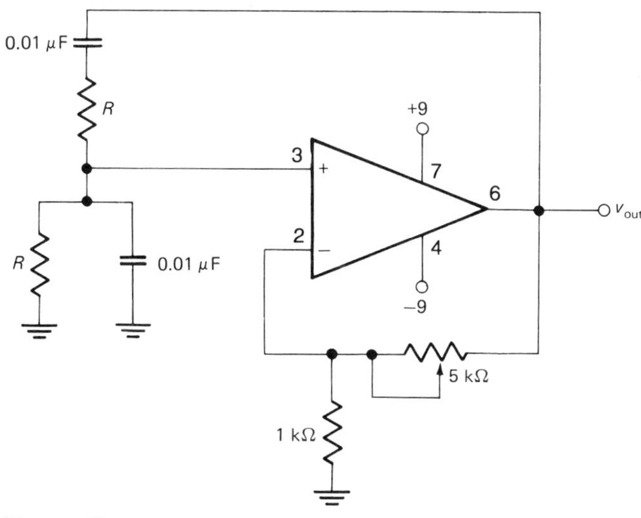

Figure 42

2. Work out oscillation frequencies for the other values of R shown in Table 1.
3. Connect the circuit of Fig. 42 with an R of 10 kΩ.
4. Look at the output with an oscilloscope. Adjust the potentiometer to get as large a sine wave as possible without excessive clipping.
5. Measure the period and record this value in Table 2. Calculate and record the corresponding frequency.
6. Repeat steps 3 through 5 for the remaining R values in Table 2.

DATA

Table 1.

R	*f*
10 kΩ	
4.7 kΩ	
2.2 kΩ	

Table 2.

R	*T*	*f*
10 kΩ		
4.7 kΩ		
2.2 kΩ		

43 *the* lc *oscillator*

REQUIRED READING

Pages 544–549

EQUIPMENT

1 power supply: 15 V
1 transistor: 2N3904 (you can substitute almost any small-signal *npn* silicon transistor)
4 ½-W resistors: 4.7 kΩ, 10 kΩ, 22 kΩ, 47 kΩ
3 capacitors: 0.001 μF, 0.01 μF, 0.1 μF
1 oscilloscope
1 inductor: 100 μH (If necessary, roll your own as follows: use a 1-W, 1-MΩ resistor as a core. Wrap 150 turns of AWG-32 to 40 wire on this core. The resulting inductor will be close enough to 100 μH for this experiment.)

PROCEDURE

1. Visualize the ac equivalent circuit of Fig. 43. Is it a CB or CE oscillator? Write the answer here: ______________________________.
2. Neglect transistor and stray-wiring capacitance. Calculate the resonant frequency of the oscillator and record here: f_0 =______________________________.
3. Connect the *LC* oscillator of Fig. 43.
4. Look at the output signal. You should have a sine wave. Measure the period: T= ____________________. Record the corresponding frequency: f_0 = ____________________.
5. Measure the peak-to-peak output and record here: V_{out} = ____________________ (unloaded).
6. Connect a 10-kΩ resistor across the output. The sine wave should decrease in size. Record the new value: V_{out} = ____________________ (loaded by 10 kΩ).

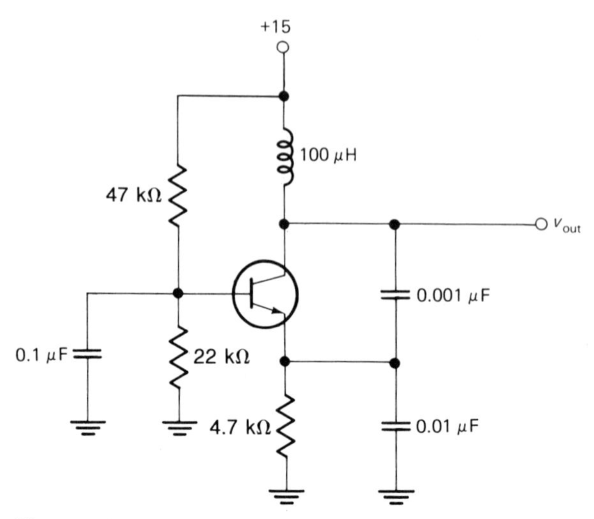

Figure 43

7. How much ac power is there in the 10-kΩ resistor? Record: P_{out} = ________.

8. Look at the signal across the 4.7-kΩ resistor. Write the peak-to-peak value here: V_e = ________________.

44 *unwanted oscillations*

THEORY

When you build a high-gain three-stage amplifier, you are likely to get unwanted oscillations unless special precautions are taken. As described in your textbook, you may get motorboating (low-frequency oscillations) because of power-supply impedance or you may get high-frequency oscillations caused by lead inductance, interstage coupling, or ground loops. In this experiment, you will build a three-stage amplifier that produces unwanted oscillations.

REQUIRED READING

Pages 553–558

EQUIPMENT

1 power supply: 15 V
3 transistors: 2N3904
7 ½-W resistors: 10 Ω, three 1 kΩ, three 220 kΩ
3 capacitors: 1 μF, two 47 μF (25-V rating or better)
1 oscilloscope
4 feet of hookup wire

PROCEDURE

1. Build the three-stage amplifier of Fig. 44.
2. Use the oscilloscope to look at the output of the amplifier. You should be getting motorboating because the 10-Ω resistor simulates an excessively high power-supply impedance. Measure and record the period of this motorboating (Table 1).
3. Replace the 10-Ω resistor by 4 feet of hookup wire. This long lead produces excessive lead inductance as described in your textbook.
4. You now should be getting high-frequency oscillations. Measure and record the period (Table 1).

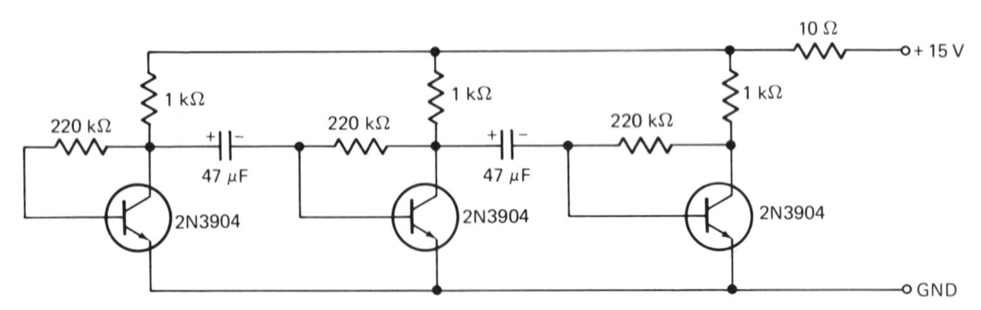

Figure 44

5. Connect a 1-μF capacitor across the supply input to the amplifier. (Refer to Fig. 19–17 in your textbook.) This may or may not eliminate the unwanted oscillations.
6. Remove the 4 feet of hookup wire and the 1-μF capacitor. Reconnect the power supply with as short a lead as possible. If oscillations are still present, try connecting the 1-μF capacitor across the supply input to the amplifier. Again, this may or may not eliminate the oscillations. If not, the oscillations may be produced by ground loops or by interstage coupling.
7. Optional: If instructor desires, try to eliminate the oscillations (if they are still present) by using a single-point ground system, by increasing the distance between stages, by shielding, etc.

DATA

Table 1. Period of Oscillations

Step 2	______________
Step 4	______________

45 the silicon controlled rectifier

REQUIRED READING

Pages 558–563

EQUIPMENT

2 power supplies (adjustable from 1 to 15 V minimum)
2 SCRs: 2N4444
1 LED: TIL222 (or any LED that can handle up to 50 mA)
2 1-W resistors: 330 Ω
1 VOM

PROCEDURE

1. Build the circuit shown in Fig. 45 with both supplies initially at 0 V.

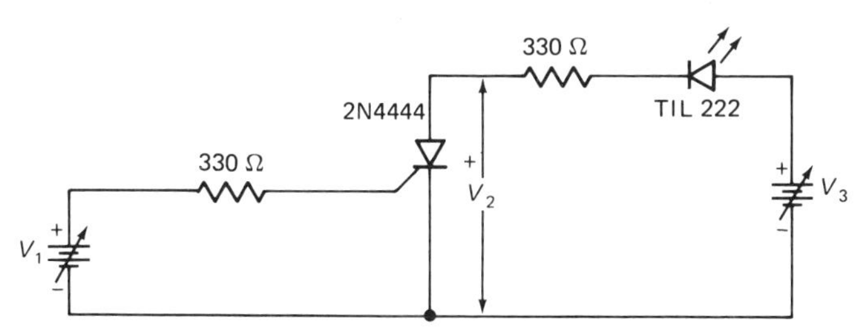

Figure 45

2. Adjust V_3 to 15 V. Slowly increase V_1 until the LED comes on. Measure voltages V_1 and V_2 and record in Table 1.
3. Reduce V_1 to zero. Does the LED go out? Write yes or no here: ______.
4. Slowly decrease V_3 until the LED just goes out. Measure and record V_3 and V_2 (Table 1).

5. Repeat steps 2 through 4 using the second SCR, recording the data in Table 2.

DATA

Table 1.

V_1	
V_2 in step 2	
V_3	
V_2 in step 4	

Table 2.

V_1	
V_2 in step 2	
V_3	
V_2 in step 4	

46 *voltage regulation*

REQUIRED READING

Pages 571–577

EQUIPMENT

1 power supply: adjustable from 15 to 25 V
1 diode: 1N753
3 transistors: 2N3904 (or almost any small-signal *npn* silicon transistor)
5 ½-W resistors: 33 Ω, 680 Ω, 2.2 kΩ, 3.3kΩ, 10 kΩ
1 potentiometer: 5 kΩ
1 decade resistance box (if unavailable, use 100-kΩ potentiometer)
1 VOM

PROCEDURE

1. In Fig. 46, what is the value of A_{SP} when the wiper is all the way up? When it is all the way down? Record the answers in Table 1.
2. If V_Z is 6.2 V, calculate the minimum and maximum load voltages. Record the values in Table 1. Use Eq. (20–6) with $V_{BE} = 0.7$ V.
3. Suppose Q_3 turns on when its V_{BE} equals 0.66 V. What is the value of load current where current limiting begins? Record as I_{max} in Table 1.
4. Connect the SP regulator of Fig. 46.
5. Set the power supply to get a V_S of 20 V.
6. Adjust the 5-kΩ potentiometer to get minimum V_L. Record this minimum load voltage in Table 2. Similarly, adjust to get maximum load voltage and record in Table 2.
7. Adjust the potentiometer to get a V_L of 10 V. Connect the decade resistance box across the load terminals. Change the resistance from 100 kΩ to 1 kΩ. As you do this, the voltage will decrease slightly. Record the change as $\Delta V_{L(\text{load})}$ in Table 2.
8. Decrease the decade resistance until the load voltage starts dropping off. Current limiting is now taking place. The lower you make the resistance, the lower the voltage will drop.

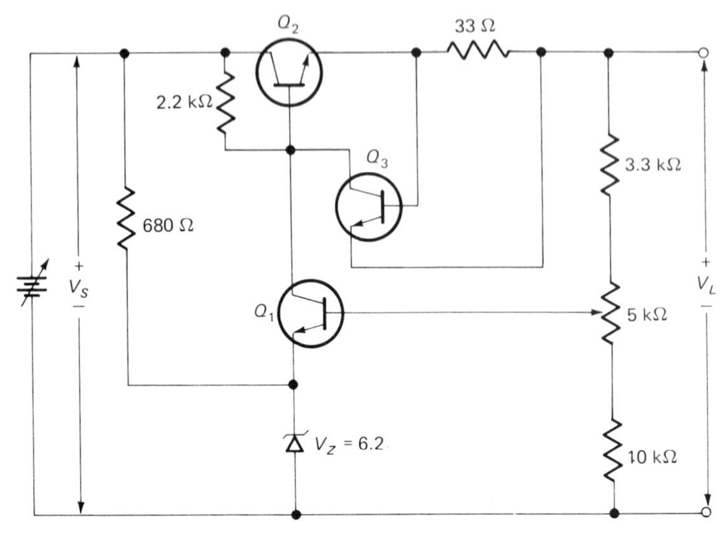

Figure 46

9. Set the decade resistance to 1 kΩ. Throw a short circuit across the load terminals and notice how the voltage drops to zero. Remove the short circuit and notice how the voltage returns to 10 V.
10. With the load voltage at 10 V, vary the input supply to change V_S from 17 to 23 V (this simulates a 20 percent increase in line voltage). The load voltage will change slightly as you do this. Record the change as $\Delta V_{L(\text{line})}$ in Table 2.
11. Set V_S at 20 V, V_L at 10 V, and decade resistance at 1 kΩ. Measure the dc voltages on each transistor terminal. Write these values on the schematic diagram (Fig. 46) next to each transistor terminal.

DATA

Table 1. Calculations

$A_{SP(\text{min})}$	______
$A_{SP(\text{max})}$	______
$V_{L(\text{min})}$	______
$V_{L(\text{max})}$	______
I_{max}	______

Table 2. Measurements

$V_{L(\text{min})}$	______
$V_{L(\text{max})}$	______
$\Delta V_{L(\text{load})}$	______
$\Delta V_{L(\text{line})}$	______

47 three-terminal ic regulators

REQUIRED READING

Pages 580–583

EQUIPMENT

1 power supply: 25 V
1 power supply: adjustable from at least 1 to 15 V
1 audio generator
1 LM340-8
4 ½-W resistors: 47 Ω, 100 Ω, 150 Ω, 180 Ω
1 capacitor: 1 μF (25-V rating or better)
1 VOM
1 oscilloscope

PROCEDURE

1. Build the circuit of Fig. 47*a*.
2. Use the dc-coupled input of the oscilloscope to look at the output of the regulator (pin 2).
3. Slowly increase the adjustable supply from 1 to 15 V and observe the oscilloscope display. If oscillations are present, connect a 1-μF capacitor across the input.
4. Connect a VOM across the regulator output.
5. Adjust the input supply to the values listed in Table 1 and record the output voltages from the regulator.
6. Build the circuit of Fig. 47*b*.
7. Use the ac-coupled input of the oscilloscope to look at the ripple input to the regulator. Adjust the generator to get 1 V peak-to-peak.
8. Use the ac-coupled input of the oscilloscope to look at the output ripple from the regulator. Increase the oscilloscope sensitivity until you can measure the peak-to-peak value. Record in Table 2.
9. Build the circuit of Fig. 47*c* with $R_2 = 47\ \Omega$.
10. Measure and record V_{OUT} in Table 3.

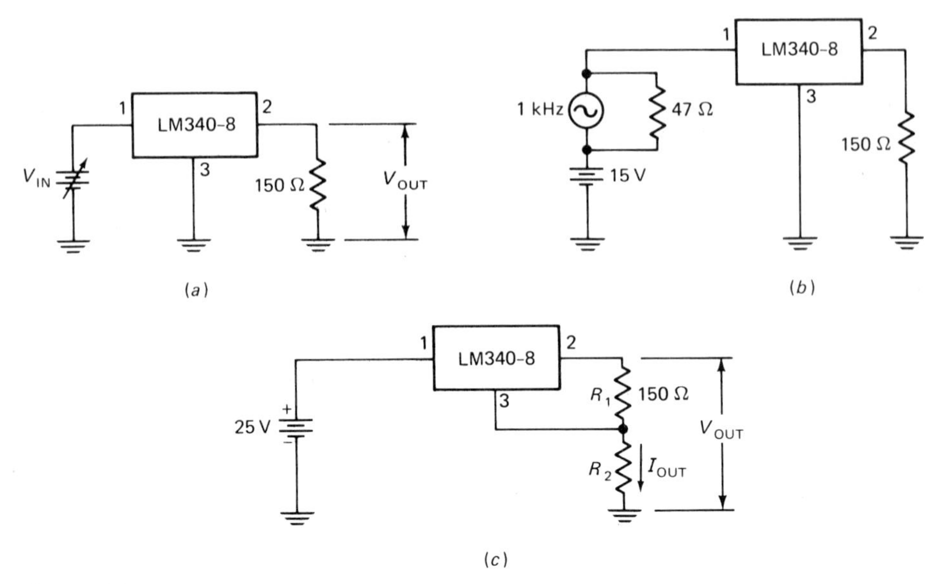

Figure 47

11. Change R_2 to the values listed in Table 3 and record V_{OUT} for each value.
12. Use the VOM to measure the current through R_2. Record the value of current for each R_2 value listed in Table 4.

DATA

Table 1.

V_{IN}	V_{OUT}
1 V	
5 V	
10 V	
11 V	
12 V	
13 V	
14 V	
15 V	

Table 2.

Output ripple	____________

Table 3.

R_2	V_{OUT}
47 Ω	
100 Ω	
180 Ω	

Table 4.

R_2	I_{OUT}
47 Ω	
100 Ω	
180 Ω	

48 *comparators and amplifiers*

REQUIRED READING

Pages 588–595

EQUIPMENT

1 audio generator
2 power supplies: 15 V
1 op amp: 741C (DIL-8 or TO-5 package)
2 LEDs: TIL221 and TIL222 (or equivalent red and green LEDs that can handle up to 50 mA)
4 ½-W resistors: 10 Ω, 100 Ω, 1 kΩ, 10 kΩ
1 potentiometer: 5 kΩ
1 oscilloscope
1 VOM

PROCEDURE

1. Build the circuit of Fig. 48*a*.
2. Vary the potentiometer and notice what the LEDs do.
3. Use the dc-coupled input of the oscilloscope to look at the input voltage to pin 3. Adjust the potentiometer to get +100 mV input. Record the color of the on LED (Table 1).
4. Adjust the potentiometer to get −100 mV. Record the color of the on LED (Table 1).
5. Build the amplifier of Fig. 48*b*.
6. Adjust the signal generator to get a 1-kHz output.
7. Use the oscilloscope to look at the output signal. Adjust the generator to get 8 V peak-to-peak.
8. Measure the peak-to-peak input voltage. Record V_{IN} in Table 2.
9. Build the circuit of Fig. 48*c* with $R = 10\ \Omega$.
10. Measure and record the current through R (Table 2).
11. Change R to 100 Ω. Measure and record the current through R (Table 2).

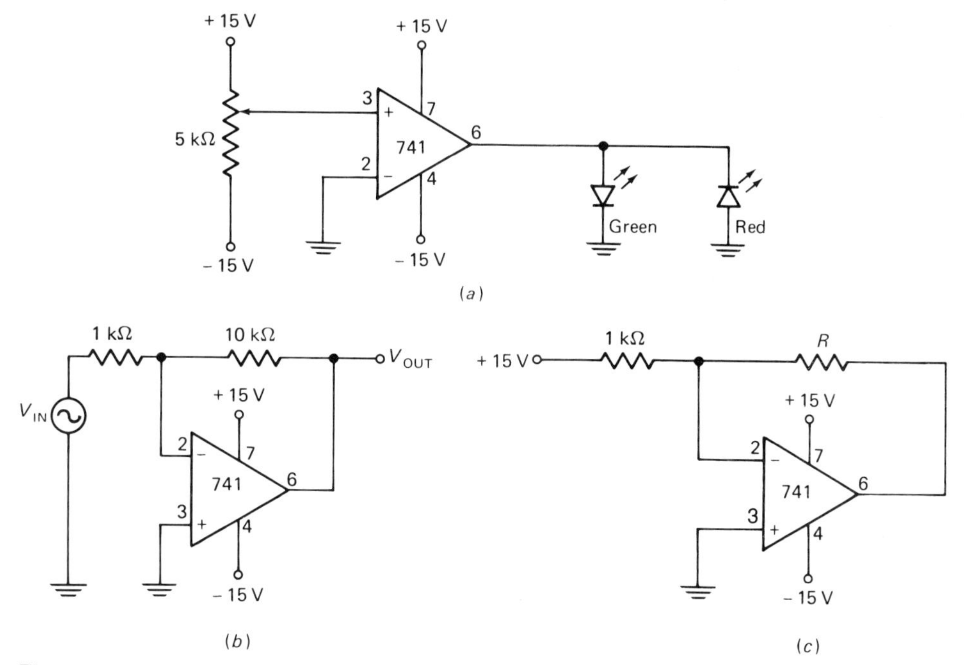

Figure 48

DATA

Table 1. Comparator

Color (step 3)	
Color (step 4)	

Table 2. Amplifier

V_{IN} (step 8)	
I_{OUT} (step 10)	
I_{OUT} (step 11)	

49 *active diode circuits*

REQUIRED READING

Pages 595–598

EQUIPMENT

1 audio generator
2 power supplies: 15 V
1 diode: 1N914
1 op amp: 741C (DIL-8 or TO-5 package)
2 ½-W resistors: 2.2 kΩ, 10 kΩ
1 potentiometer: 5 kΩ
1 oscilloscope

PROCEDURE

1. Build the half-wave rectifier of Fig. 49 *a*.
2. Connect the oscilloscope (dc input) across the 10-kΩ load resistor.
3. Set the generator to 100 Hz and adjust the level to get a 1-V peak output on the oscilloscope.
4. If dual-beam or dual-trace oscilloscope is being used, use the other input to look at the input signal (pin 3). If single-trace oscilloscope is being used, connect it to the input (pin 3). Record the peak value of the input sine wave (Table 1).
5. Adjust the signal level to get a half-wave rectified output with a peak value of 100 mV. Measure and record the input peak value (Table 1).
6. Connect a 47-μF capacitor across the load to get the circuit of Fig. 49 *b*.
7. Adjust the generator to get an input peak value of 1 V. Measure and record the dc output value (Table 2).
8. Readjust the generator to get an input peak value of 100 mV. Measure and record the dc output (Table 2).
9. Build the circuit of Fig. 49 *c*.
10. Adjust the generator to produce a peak value of 1 V at the left end of the 2.2-kΩ resistor.

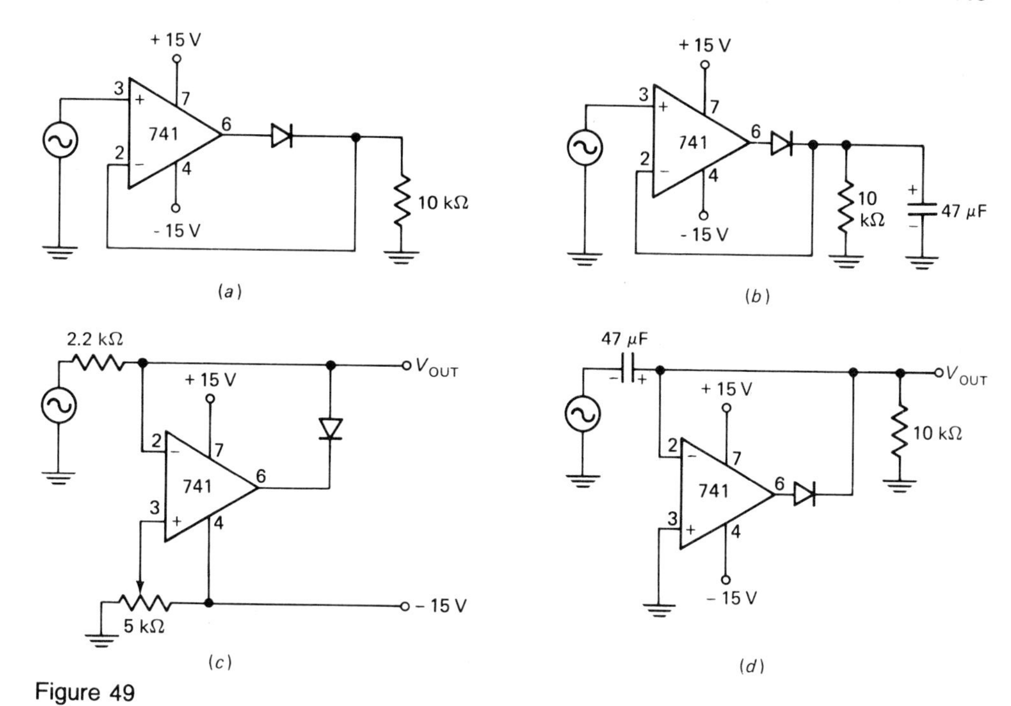

Figure 49

11. Look at the output signal while turning the potentiometer through its entire range.
12. Adjust the generator to get a peak value of 100 mV at the left end of the 2.2-kΩ resistor. Repeat step 11. Reverse the diode polarity and repeat step 11.
13. Build the circuit shown in Fig. 49*d*.
14. Look at the output using the dc-coupled input of the oscilloscope. Set the generator to 100 Hz and adjust the level to produce a 2-V peak-to-peak signal. Be sure to note where the dc ground level is.
15. Look at the input signal (left end of capacitor) and confirm that it is approximately 2 V peak-to-peak.

DATA

Table 1. Half-Wave Rectifier

V_P (step 4)	______
V_P (step 5)	______

Table 2. Peak Detector

V_{DC} (step 7)	______
V_{DC} (step 8)	______

50 *active filters*

REQUIRED READING

Pages 604–607

EQUIPMENT

1 audio generator
2 power supplies: 15 V
1 op amp: 741C (DIL-8 or TO-5 package)
3 ½-W resistors: two 10 kΩ, 20 kΩ
3 capacitors: two 0.01 μF, 0.02 μF
1 oscilloscope
1 ac millivoltmeter

PROCEDURE

1. Build the circuit of Fig. 50*a*.
2. Set the input frequency to 100 Hz. Adjust the level to get 2 V peak-to-peak at the output.
3. If a dual-trace oscilloscope is available, use the other input to look at the V_{IN} signal. If a single-trace oscilloscope is being used, connect it across the generator. Measure and record the peak-to-peak input signal (Table 1).
4. Increase the frequency until the output signal breaks or corners (down 3 dB). Use the ac millivoltmeter to locate the exact 3-dB corner frequency. Record the break frequency in Table 1.
5. Increase the frequency to 10 times the break frequency. Measure and record V_{IN} and V_{OUT} (Table 1).
6. Build the circuit of Fig. 50*b*.
7. Set the frequency to 10 kHz and adjust the level to get 1 V rms at the output. Use the oscilloscope to confirm the presence of a sine-wave signal.
8. Look at the input with the oscilloscope. It should be a sine wave with the same peak value as the output.

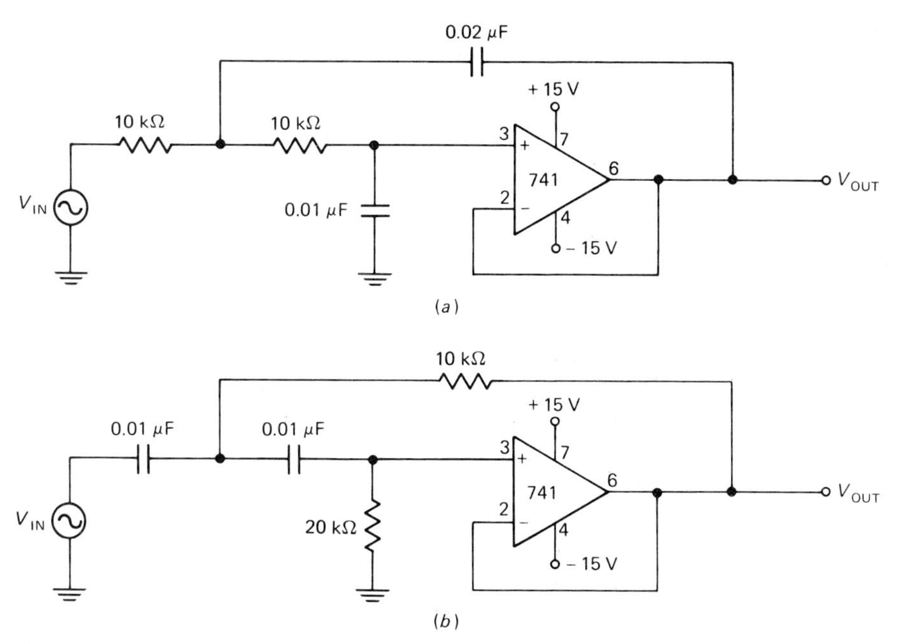

Figure 50

9. Locate the approximate break frequency using the oscilloscope. Then, use the ac millivoltmeter to find the exact break frequency. Record this frequency in Table 2.
10. Decrease the input frequency to one-tenth of the break frequency. Measure and record V_{IN} and V_{OUT} (Table 2).

DATA

Table 1. Low-Pass Filter

V_{IN} (step 3)	
Corner frequency	
V_{IN} (step 5)	
V_{OUT} (step 5)	

Table 2. High-Pass Filter

Corner frequency	______
V_{IN} (step 10)	______
V_{OUT} (step 10)	______

51 *harmonic distortion*

REQUIRED READING

Pages 629–636

EQUIPMENT

1 audio generator
1 power supply: 25 V
1 JFET: MPF102 or any substitute with an I_{DSS} greater than 2 mA
6 ½-W resistors: 2.2 kΩ, 4.7 kΩ, 47 kΩ, two 100 kΩ, 220 kΩ
4 capacitors: two 1000 pF, 2000 pF, 10 μF
1 ac millivoltmeter
1 oscilloscope

PROCEDURE

1. Calculate the notch frequency of the twin-T filter shown in Fig. 51. Write the value here: f_0 = ________________________.

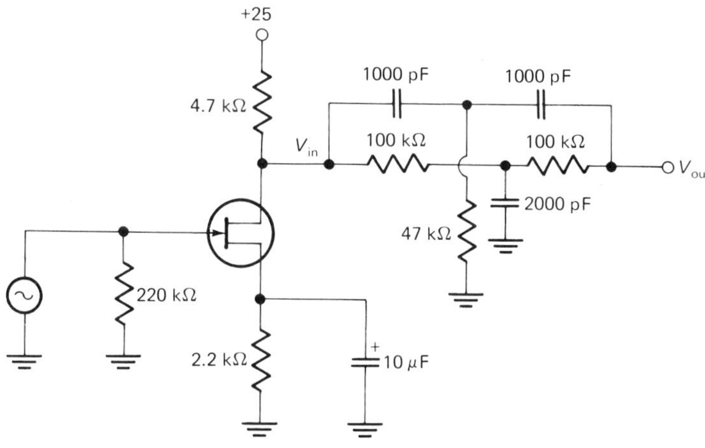

Figure 51

2. Connect the circuit of Fig. 51.
3. Set the frequency of the audio generator at the f_0 calculated in step 1.
4. Connect the ac millivoltmeter and the oscilloscope across the input to the notch filter (the drain-to-ground terminals). Set the signal level to get a V_{in} of 1 V rms to the notch filter. The waveform on the oscilloscope should look like a sine wave.
5. Connect the ac millivoltmeter and the oscilloscope across the output of the notch filter. Increase the vertical sensitivity by a factor of 10. Vary the frequency until the ac millivoltmeter reads minimum. The final frequency is the true notch frequency. It should be approximately the same as that calculated in step 1. Record the true notch frequency here:

$$f_0 = \_\_\_\_\_\_\_\_\_\_\_\_$$

6. Vary the frequency above and below the true notch frequency. Notice how the waveform on the oscilloscope changes as you notch out the fundamental.
7. Set the frequency to get a minimum reading on the ac millivoltmeter. Record the value in Table 1 under V_{out}.
8. Calculate the approximate value[1] of harmonic distortion using

$$\% \text{ harm dist} = \frac{V_{out}}{V_{in}} \times 100\%$$

   Record the answer in Table 1.
9. Repeat steps 4 through 8 for the remaining V_{in} values of Table 1.
10. Connect the oscilloscope across the input to the notch filter. Set the signal level to get 1 V pp. Adjust the sweep speed to get a couple of cycles on the screen. Notice how the waveform looks almost like a perfect sine wave.
11. Increase the signal level until clipping just occurs on the upper peaks. Reduce the level slightly. If you look closely, you will see the lower half cycles get more gain than the upper half cycles. In other words, you get nonlinear distortion similar to Fig. 22–22 (textbook) except that the wave is inverted.

[1] This is an approximation for several reasons. If interested, see Malvino, A. P., "Electronic Instrumentation Fundamentals," pp. 345–351, McGraw-Hill Book Company, New York, 1967.

DATA

Table 1. Distortion Measurements

V_{in}	V_{out}	% harm dist
1 V		
2 V		
3 V		
4 V		
5 V		

52 the frequency mixer

REQUIRED READING

Pages 668–673

EQUIPMENT

2 audio generators
1 power supply: 25 V
1 transistor: 2N3904 (or almost any small-signal *npn* silicon transistor)
9 ½-W resistors: 4.7 kΩ, two 10 kΩ, two 15 kΩ, 22 kΩ, two 33 kΩ, 68 kΩ
7 capacitors: two 100 pF, two 220 pF, 1000 pF, 0.1 μF, 1 μF
1 oscilloscope

PROCEDURE

1. Figure 52 shows a bipolar mixer where v_x is the small signal and v_y the large signal. If f_x is 51 kHz and f_y is 50 kHz, write the following group-1 frequencies:

$f_x \pm f_y =$ ________________ & ________________

$f_x \pm 2f_y =$ ________________ & ________________

$f_x \pm 3f_y =$ ________________ & ________________

2. Write the following group-2 frequencies:

$2f_x \pm f_y =$ ________________ & ________________

$2f_x \pm 2f_y =$ ________________ & ________________

$2f_x \pm 3f_y =$ ________________ & ________________

3. Write the following group-3 frequencies:

$3f_x \pm f_y =$ ________________ & ________________

$3f_x \pm 2f_y =$ ________________ & ________________

$3f_x \pm 3f_y =$ ________________ & ________________

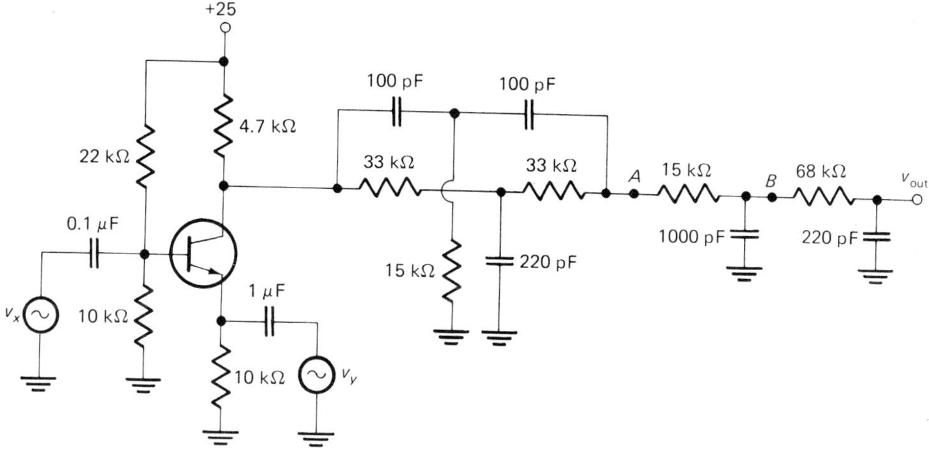

Figure 52

4. All the foregoing frequencies plus other intermodulation components will appear at the collector of Fig. 52. The twin-T filter has a notch of approximately 50 kHz, so it will greatly attenuate f_x and f_y. In addition, each lag network has a break frequency of approximately 10 kHz. Ideally, if the filters stop all frequencies above 10 kHz, which of the frequencies recorded in steps 1, 2, and 3 appear at the final output?

   Write the answers here: ______________, ______________,

   and ______________.
5. Connect the circuit of Fig. 52.
6. Turn v_y down to zero. With the oscilloscope, adjust v_x to 0.1 V pp. Set the frequency at 51 kHz (approximately).
7. Next adjust v_y to 2 V pp and 50 kHz.
8. Look at the final output signal with the vertical sensitivity at 0.1 V/cm (ac input) and sweep time at 0.2 ms/cm. Vary the frequency of the v_x generator slowly in the vicinity of 51 kHz until you get a 1-kHz output signal.
9. Look at point *B*, the input to the final lag network. Notice the ripple on the 1-kHz signal.
10. Look at point *A*, the input to the first lag network. Notice how large the ripple is here.
11. Return to the final output. If the frequency has drifted, readjust f_x to get a 1-kHz output. Record the peak-to-peak voltage here:

$$V_{out} = \underline{\hspace{3cm}}$$

53 amplitude modulation

REQUIRED READING

Pages 686–695

EQUIPMENT

1 audio generator
1 RF generator
1 power supply: 25 V
1 transistor: 2N3904 (or almost any small-signal *npn* silicon transistor)
5 ½-W resistors: 1 kΩ, three 10 kΩ, 22 kΩ
3 capacitors: 0.001 μF, 0.01 μF, 0.1 μF
1 oscilloscope

PROCEDURE

1. If the audio signal is zero and all capacitors look like high-frequency shorts, what is the ideal voltage gain for the carrier in Fig. 53? Write the value here: A_0 = ________________.
2. If the audio signal has a peak-to-peak value of 10 V, what are the minimum and maximum voltage gains for the carrier?

   A_{max} = ________________ and A_{min} = ________________.
3. With the foregoing gains, what is the percent modulation? Record the answer here: ________________.
4. Connect the circuit of Fig. 53.
5. Set the audio generator to 200 Hz and the RF generator to 500 kHz.
6. Turn the audio generator down to zero (do not disconnect). Adjust the RF generator to get a final output v_{out} of 0.3 V pp (unmodulated).
7. Measure the RF input signal v_x and record the peak-to-peak value here: v_x = ________________. Calculate the value of A_0 and record here: A_0 = ________________.

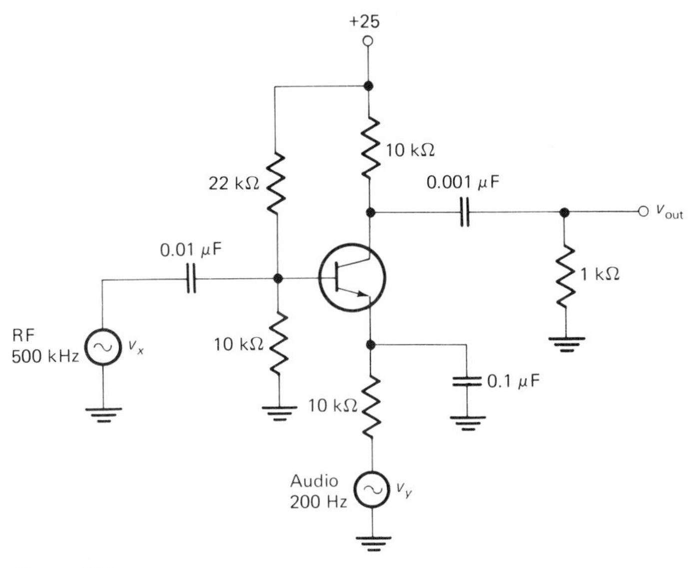

Figure 53

8. Look at v_{out} with the sweep speed at 1 ms/cm. Turn up the audio signal and you will see amplitude modulation.
9. Look at v_y and set the level at 5 V pp.
10. Look at v_{out} and record the maximum and minimum peak-to-peak values: $2V_{max}$ = ________________ and $2V_{min}$ = ________________. Calculate the percent modulation: ________________.
11. Increase and decrease the audio level and notice how the percent modulation changes.

54 *the envelope detector*

REQUIRED READING

Pages 704–706

EQUIPMENT

1 RF generator
1 diode: 1N914 (or almost any small-signal diode)
1 transistor: 2N3904 (or almost any small-signal *npn* silicon transistor)
4 ½-W resistors: two 1 kΩ, 10 kΩ, 100 kΩ
3 capacitors: two 0.001 μF, 0.01 μF
1 oscilloscope

PROCEDURE

1. Work out the highest frequency the envelope detector of Fig. 54 can follow without attenuation for a modulation of 30 percent. Record here: $f_{y(\max)}$ = ________________.

2. Connect the circuit of Fig. 54.

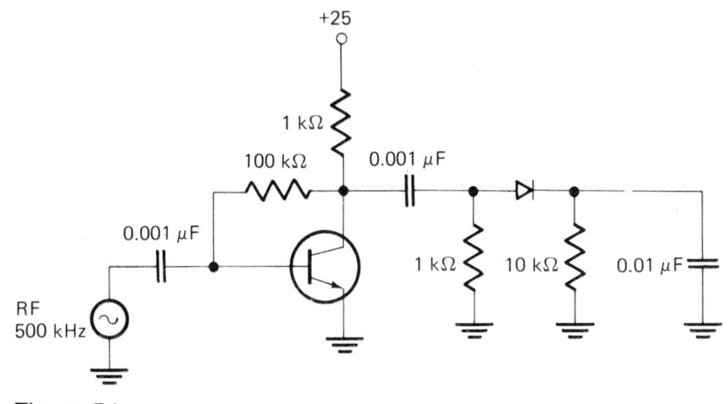

Figure 54

3. Set the RF generator to 500 kHz unmodulated. Look at the input to the envelope detector (across the lower 1-kΩ resistor). Adjust the RF generator to get a signal level of 5 V pp with the sweep time at 1 ms/cm.
4. Turn on the modulation. If it is adjustable, set the modulation at 30 percent.
5. Look at the output of the envelope detector. You should have an audio signal.

Name: TED LAIL Date: 1-8-82

5/5

QUESTIONS FOR EXPERIMENT 1

1. The data of Table 1 prove load voltage is:
   (a) perfectly constant; *(b)* small; *(c)* heavily dependent on load resistance; *(d)* approximately constant. (d)
2. When internal resistance R increases in Fig. 1*a*, load voltage:
   (a) increases; *(b)* decreases; *(c)* stays the same. (b)
3. In Fig. 1*a*, you can neglect internal resistance with less than 1% error provided R is less than:
   (a) 10 Ω; *(b)* 100 Ω; *(c)* 500 Ω; *(d)* 1 kΩ. (b)
4. The circuit left of the *AB* terminals in Fig. 1*b* acts approximately like a current source because the current values of Table 2:
   (a) increases slightly; *(b)* are almost constant; *(c)* decrease a great deal; *(d)* depend heavily on load resistance. (b)
5. In Fig. 1*b*, the load resistance has little effect on current as long as the load resistance is:
   (a) much smaller than 1 kΩ; *(b)* large; *(c)* much larger than 1 kΩ; *(d)* greater than 1 kΩ. (a)
6. A car battery is closely approximated by a:
   (a) current source; *(b)* voltage source; *(c)* resistance. (b)
7. Measuring voltage out of a car battery without a load is a poor test because:
   (a) current sources need to be loaded; *(b)* internal resistance may be zero; *(c)* internal resistance may be high; *(d)* you only get voltage when a load resistance is connected. (c)
8. Show the values of Tables 1 and 2.

Table 1:

V_L 10v, 10v, 9.91, 9.55

Table 2:

I 10mA, 9.78mA, 9.51mA, 9.06mA

9. Optional: Instructor's question.

THE RESISTANCE of THE ELEMENTS & COMPONENTS <u>INSIDE</u> A POWER SUPPLY.

10. Optional: Instructor's question.

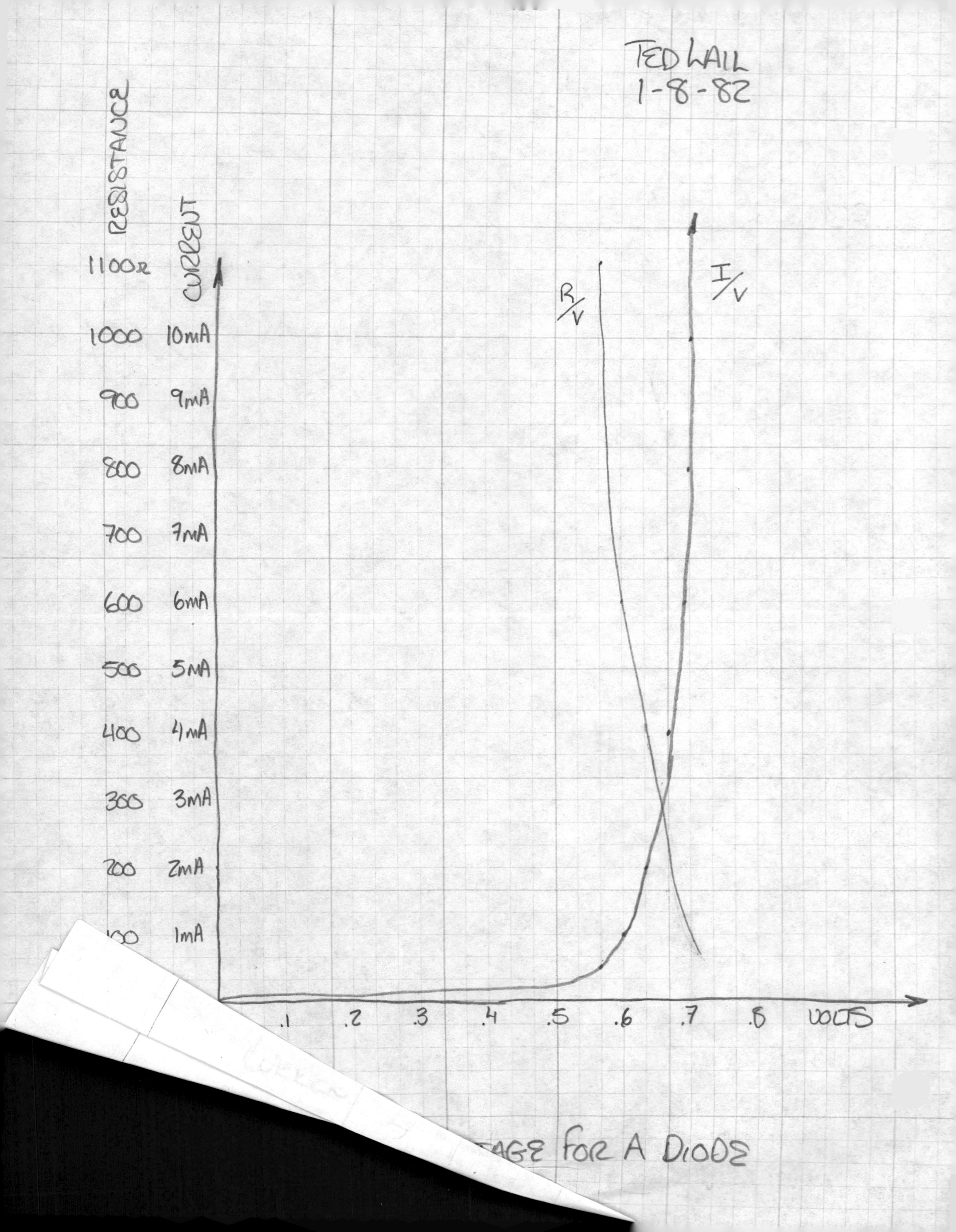
TED WALL
1-8-82
RESISTANCE
CURRENT
1100Ω
1000 10mA
900 9mA
800 8mA
700 7mA
600 6mA
500 5mA
400 4mA
300 3mA
200 2mA
1mA
R/V
I/V
.1 .2 .3 .4 .5 .6 .7 .8 VOLTS
AGE FOR A DIODE

Name: TED LAIL Date: 1-8-82

QUESTIONS FOR EXPERIMENT 2

1. In this experiment you measure Thevenin voltage with:
   (a) an ohmmeter; *(b)* the load disconnected; *(c)* the load in circuit. (b)
2. You first measured R_{TH} with a:
   (a) voltmeter; *(b)* load; *(c)* shorted source. (C)
3. You can also measure R_{TH} by the matched-load method, which involves:
   (a) an open voltage source; *(b)* a load that is open; *(c)* varying Thevenin resistance until it matches load resistance; *(d)* changing load resistance until load voltage drops to ½ V_{TH}. (d)
4. Discrepancies between calculated and measured values in Table 1 cannot be caused by:
   (a) instrument error; *(b)* resistor tolerance; *(c)* human error; *(d)* load resistance. (d)
5. If a black box puts out a constant voltage for all load resistances, the Thevenin resistance of this box approaches:
   (a) zero; *(b)* infinity; *(c)* load resistance. (a)
6. If a black box puts out a constant load current for all load resistances, the Thevenin resistance is ideally:
   (a) zero; *(b)* infinite; *(c)* equal to load resistance. (b)
7. Measuring the voltage out of a car battery under open-load conditions is a poor test because:
   (a) R_{TH} may be too low; *(b)* R_{TH} may be infinite; *(c)* R_{TH} may be large enough to drop voltage under load. (C)
8. Ideally, a voltmeter should have infinite resistance. Explain why the voltmeter introduced a small error in step 5 of the Procedure.

   THE ERROR WAS ONLY .9%
   IF I HAD USED A VOM IT WOULD
   HAVE BEEN MUCH GREATER.

9. Optional: Instructor's question.

10. Optional: Instructor's question.

Name: TED LAIL **Date:** 1-8-82

QUESTIONS FOR EXPERIMENT 3

1. You can find diode current in Fig. 3*c* without an ammeter by applying Ohm's law to the:
   (a) source; *(b)* diode; *(c)* load; *(d)* entire circuit. (C)
2. In this experiment the knee voltage is closest to:
   (a) 0.3 V; *(b)* 0.7 V; *(c)* 1 V; *(d)* 1.2 V. (b)
3. With forward bias, the dc resistance increases when:
   (a) current increases; *(b)* diode voltage increases; *(c)* the ratio of v/i decreases; *(d)* the ratio of i/v decreases. (d)
4. In Table 2, you cannot calculate the first R value because:
   (a) of 0/0; *(b)* anything with zero current must have infinite resistance; *(c)* anything with zero voltage must have zero current; *(d)* it's zero. (C) 0 Ω
5. A diode acts like a high resistance at:
   (a) low currents; *(b)* medium currents; *(c)* high currents; *(d)* high voltages. (a)
6. Which of the following is the strongest reason for diode current being equal to load current in Fig. 3*b*?
   (a) Ohm's law; *(b)* Thevenin's theorem; *(c)* Kirchhoff's current law; *(d)* Kirchhoff's voltage law. (C)
7. Which of the following most closely describes the diode curve above the knee?
   (a) It becomes horizontal; *(b)* voltage increases rapidly; *(c)* current increases rapidly; *(d)* dc resistance increases rapidly. (C)
8. Write the R values of Table 2 here:

V_L	R
0.5	1.14 kΩ
1	600 Ω
2	320 Ω
4	167.5 Ω
6	115 Ω
8	87.5 Ω
10	71 Ω

9. Optional: Instructor's question.

10. Optional: Instructor's question.

Name: TED LAIL Date: 1-11-82

QUESTIONS FOR EXPERIMENT 4

1. In this experiment, knee voltage is the diode voltage that: *(a)* equals 0.3 V; *(b)* equals 0.7 V; *(c)* corresponds to 10 mA; *(d)* corresponds to 50 mA. (C)
2. Bulk resistance is not: *(a)* a ratio; *(b)* the voltage difference divided by the current difference of two points above the knee; *(c)* in ohms; *(d)* the same as dc resistance. (d)
3. The dc resistance of a silicon diode for a current of 10 mA is closest to: *(a)* 2.5 Ω; *(b)* 10 Ω; *(c)* 70 Ω; *(d)* 1 kΩ. C ~~(a)~~ C
4. In Fig. 4*b*, the power dissipated by the diode equals the product of voltage and current. This power is closest to: *(a)* 0; *(b)* 1.5 mW; *(c)* 15 mW; *(d)* 150 mW. (C)
5. If you neglect knee voltage, the error in Fig. 4*b* is: *(a)* less than 10%; *(b)* more than 10%; *(c)* less than 1%. (a)
6. If you neglect bulk resistance in Fig. 4*b*, the error is: *(a)* less than 10%; *(b)* more than 10%. (a)
7. Suppose the diode of Fig. 4*b* has an $I_{F(max)}$ of 500 mA. To avoid destroying the diode, the source voltage can be no more than: *(a)* 15 V; *(b)* 50 V; *(c)* 185 V; *(d)* 272 V. (d)
8. Record the values of Tables 2 and 3 here:

V_{knee} .727 V

r_B 2.35 Ω

Experimental I 26.0 mA

Ideal I 27.6 mA

Second I 25.6 mA

Third I 25.5 mA

9. Optional: Instructor's question.

10. Optional: Instructor's question.

Name: TED LAIL Date: 1-11-82

QUESTIONS FOR EXPERIMENT 5

1. In Fig. 5, the zener current and the current through the 180-Ω resistor are:
   (a) equal; *(b)* almost equal; *(c)* much different. (a)
2. The zener diode starts to break down when the input voltage is approximately:
   (a) 4 V; *(b)* 6 V; *(c)* 8 V; *(d)* 10 V. (b)
3. When V_{IN} is less than 6 V, the output voltage is:
   (a) approximately constant; *(b)* negative; *(c)* the same as the input. (C)
4. When V_{IN} is greater than 8 V, the output voltage is:
   (a) approximately constant; *(b)* negative; *(c)* the same as the input. (a)
5. The zener current is closest to the specified test current when V_{IN} is:
   (a) 6 V; *(b)* 8 V; *(c)* 10 V; *(d)* 12 V. ( ) → 15V
6. The calculated zener impedance is closest to:
   (a) 1 Ω; *(b)* 2 Ω; *(c)* 7 Ω; *(d)* 20 Ω. (a)
7. In Table 1, the change in the output voltage as the input voltage increases from 10 to 15 V is closest to: (.05)
   (a) 0.01 V; *(b)* 0.1 V; *(c)* 1 V; *(d)* 100 μV. (b)
8. The maximum allowable zener current for a 1N753 is closest to:
   (a) 20 mA; *(b)* 40 mA; *(c)* 60 mA; *(d)* 90 mA. ( ) → 161 mA

9. Optional: Instructor's question.

10. Optional: Instructor's question.

Name: TED LAIL Date: 1-13-82

SKS

QUESTIONS FOR EXPERIMENT 6

1. To measure the rms voltage across the secondary winding in Fig. 6*a*, it's best to use:
   (a) a VOM; *(b)* an oscilloscope; *(c)* an ammeter. (a)
2. With the half-wave rectifier of Fig. 6*a*, the peak voltage across the load is closest to:
   (a) 5 V; *(b)* 10 V; *(c)* 20 V; *(d)* 40 V. (C)
3. The period of the half-wave signal in Fig. 6*a* is nearest to:
   (a) 8 ms; *(b)* 16 ms; *(c)* 24 ms; *(d)* 30 ms. (b)
4. With the center-tap rectifier of Fig. 6*b*, the dc load voltage is closest to:
   (a) 2 V; *(b)* 4 V; *(c)* 6 V; *(d)* 8 V. (C)
5. The period of the full-wave signal in Fig. 6*b* is approximately:
   (a) 2 ms; *(b)* 4 ms; *(c)* 8 ms; *(d)* 16 ms. (C)
6. The peak voltage out of the bridge rectifier of Fig. 6*c* compared to the peak voltage out of the center-tap rectifier of Fig. 6*b* is about:
   (a) half as large; *(b)* the same; *(c)* twice as large. (C)
7. In all rectifiers of this experiment, the dc load voltages measured slightly lower than theoretical values because of:
   (a) diode drops; *(b)* small load resistance; *(c)* low line voltage; *(d)* diode breakdown. (a)
8. In this experiment, the rectifier circuit with the largest dc output voltage is the:
   (a) half-wave rectifier; *(b)* center-tap rectifier; *(c)* bridge rectifier. (C)

9. Optional: Instructor's question.

10. Optional: Instructor's question.

Name: TED LAIL Date: 1-14-82

SKS

QUESTIONS FOR EXPERIMENT 7

1. When $R_{SURGE} = 0$, the Thevenin resistance facing the filter capacitor is closest to:
   (a) 0.1 Ω; *(b)* 1 Ω; *(c)* 20 Ω; *(d)* 220 Ω. (b)
2. In step 3, the output waveform is:
   (a) half-wave rectified; *(b)* full-wave rectified; *(c)* peak rectified; *(d)* positively clamped. (b)
3. The discharging time constant in step 4 is:
   (a) 1 ms; *(b)* 10 ms; *(c)* 100 ms; *(d)* 4700 ms. (b)
4. The ripple decreases when the filter capacitor:
   (a) decreases; *(b)* increases; *(c)* is removed. (b)
5. In step 5, the output waveform is:
   (a) half-wave rectified; *(b)* full-wave rectified; *(c)* peak rectified; *(d)* positively clamped. (C)
6. When $R_{SURGE} = 220\ \Omega$, the Thevenin resistance facing the filter capacitor is approximately:
   (a) 0.1 Ω; *(b)* 1 Ω; *(c)* 20 Ω; *(d)* 220 Ω. (d)
7. In this experiment, V_{TH} is closest to:
   (a) 12 V; *(b)* 15 V; *(c)* 20 V; *(d)* 30 V. (C)
8. In step 7, R_{TH}/R_L approximately equals:
   (a) 0; *(b)* 0.01; *(c)* 0.02; *(d)* 0.04. (C)

9. Optional: Instructor's question.

10. Optional: Instructor's question.

Name: TED LAIH Date: 1-14-82

SKS

QUESTIONS FOR EPERIMENT 8

1. The calculated peak voltage across half the secondary winding is nearest to:
   (a) 5 V; *(b)* 10 V; *(c)* 20 V; *(d)* 30 V. (b)
2. Which of these has the best voltage regulation?
   (a) ×1 output; *(b)* doubler output; *(c)* tripler output. (a)
3. The percent voltage regulation for the tripler output is closest to:
   (a) 1%; *(b)* 5%; *(c)* 10%; *(d)* 20%. (d)
4. When the 6.8-kΩ load is connected across any output, the ripple:
   (a) decreases; *(b)* stays the same; *(c)* increases. (c)
5. The ripple factor for the tripler output is closest to:
   (a) 1%; *(b)* 4%; *(c)* 12%; *(d)* 25%. 6.9% (b) (b)
6. The voltage regulation of the doubler is nearest to:
   (a) 1%; *(b)* 4%; *(c)* 8%; *(d)* 16%. (c)
7. The *VR* of the ×1 output is closest to:
   (a) 1%; *(b)* 4%; *(c)* 8%; *(d)* 16%. (b)
8. The main disadvantage of voltage multipliers is:
   (a) the number of diodes required; *(b)* the large voltage ratings of the capacitors; *(c)* regulation gets worse for higher multiplication; *(d)* the large inductors. (c)

9. Optional: Instructor's question.

10. Optional: Instructor's question.

Name: TED LAIL Date: 1-18-82

QUESTIONS FOR EXPERIMENT 9

1. A split supply has:
   (a) only one output voltage; *(b)* only a positive output voltage; *(c)* only a negative output voltage; *(d)* positive and negative outputs. (d)
2. The value of $+V_{IN}$ is closest to:
   (a) 5 V; *(b)* 10 V; *(c)* 15 V; *(d)* 20 V. (b)
3. The regulated positive output voltage is approximately:
   (a) 5.6 V; *(b)* 6.2 V; *(c)* 6.8 V; *(d)* 9 V. (b)
4. The positive zener regulator attenuates the ripple by:
   (a) a factor of 5; *(b)* more than a factor of 10; *(c)* more than 100; *(d)* unity. (b)
5. The percent VR of the positive regulator is closest to:
   (a) 1%; *(b)* 2.5%; *(c)* 5%; *(d)* 7.5%. (c)
6. If the polarity of the upper zener diode is reversed in Fig. 9, $+V_{OUT}$ becomes:
   (a) a negative voltage; *(b)* approximately 0.7 V; *(c)* about 6.2 V positive; *(d)* about 9.5 V. (b)
7. With R_L disconnected, the current through the zener diode is closest to:
   (a) 20 mA; *(b)* 30 mA; *(c)* 50 mA; *(d)* 200 mA. (a)
8. With $R_L = 470\ \Omega$, the zener current is nearest to:
   (a) 1 mA; *(b)* 6 mA; *(c)* 10 mA; *(d)* 20 mA. (c)

$$I_Z = I_S - I_L = 22.4 - 13.3 = 9.1\text{ mA}$$

$$I_S = \frac{V_{IN} - V_{OUT}}{R_S} = \frac{9.63\text{ V} - 6.27\text{ V}}{150\ \Omega} = 22.4\text{ mA}$$

$$I_L = \frac{V_{OUT}}{R_L} = \frac{6.27}{470} = 13.3\text{ mA}$$

OK

9. Optional: Instructor's question.

10. Optional: Instructor's question.

Name: TED LAIL Date: 1-20-82

5/5

QUESTIONS FOR EXPERIMENT 10

1. In a negative clipper, which of these is largest?
   (a) Positive peak; *(b)* negative peak; *(c)* knee voltage. (a)
2. The combination clipper of Fig. 10*b*:
   (a) puts out a small sine wave; *(b)* generates a small squarish wave; *(c)* has an adjustable clipping level; *(d)* has an output proportional to input. (b
3. If the dc source of Fig. 10*c* varies from 0 to 15 V, the output positive peak varies from roughly:
   (a) 0 to $V_p/2$; *(b)* 0 to V_p; *(c)* 0 to $2V_p$. (b)
4. In the combination clipper of Fig. 10*b*, which approximation is the most reasonable compromise?
   (a) Ideal; *(b)* second; *(c)* third. (b)
5. If you use the second approximation for the diodes of Fig. 10*b*, which of these statements is false?
   (a) Diodes never conduct at the same time; *(b)* the 10-kΩ resistor always has some current; *(c)* the 10-kΩ resistor has no current when diodes are off; *(d)* peak-to-peak output voltage is less than 2 V. (c) b
6. If the 100-kΩ resistor were changed to a 10-kΩ resistor in Fig. 10*a*, the positive output peak would be approximately:
   (a) 0.7 V; *(b)* 5 V; *(c)* 10 V. (a)
7. If germanium diodes are used in Fig. 10*b*, the peak-to-peak output voltage would be approximately:
   (a) 0.3 V; *(b)* 0.6 V; *(c)* 1.4 V (b
8. In Table 2, the ratio of v_{rip}/V_{dc} is closest to:
   (a) 0.1%; *(b)* 1%; *(c)* 5%; *(d)* 10%. (b

9. Optional: Instructor's question.

10. Optional: Instructor's question.

Name: TED LAIL Date: 1-22-81

5/5

QUESTIONS FOR EXPERIMENT 11

1. If the diode of Fig. 11*a* is reversed, the output will be:
   (a) positively clamped; *(b)* negatively clamped; *(c)* half-wave rectified; *(d)* peak rectified. (b)
2. If V_p is 10 V in Fig. 11*b*, the peak voltage across the first diode is ideally:
   (a) 5 V; *(b)* 10 V; *(c)* 15 V; *(d)* 20 V. (d)
3. If V_p is 10 V in Fig. 11*b*, the dc output voltage is ideally:
   (a) 5 V; *(b)* 10 V; *(c)* 15 V; *(d)* 20 V. (d)
4. In Table 2, the ratio of v_{rip}/V_{dc} is closest to:
   (a) 0.1%; *(b)* 1%; *(c)* 5%; *(d)* 10%. (b)
5. When the dc return of Fig. 11*c* is disconnected, which of the following is false?
   (a) The capacitor charges to approximately V_p; *(b)* capacitor current can only be in one direction with an ideal diode; *(c)* the diode shuts off after capacitor voltage reaches maximum; *(d)* the diode becomes forward-biased. (d)
6. The way to change a positive clamper to a negative clamper is by reversing the:
   (a) source; *(b)* capacitor; *(c)* diode; *(d)* load. +6 (c)
7. Show the values of Tables 1 and 2 here:
   - V_{pp} 20v
   - out post peak 20v
   - out neg peak 0v
   - v_{rip} 150mV
   - V_{dc} 17.54v
8. Explain the main ideas behind clamping action.

BY PUTTING A DIODE IN PARALLEL W/ R_L AND IN SERIES WITH A CAPACITOR AND SOURCE THE CAPACITOR WILL CHARGE UP TO V_P WHEN THE DIODE IS FORWARD BIASED ~~AND DISCHARGE THROUGH R_L~~ TOGETHER W/ ~~SOURCE THEREBY DOUBLING V_{OUT} THRU R_L~~. THE LOWEST VOLTAGE WILL BE ZERO (-.7v) IN THIS SINE WAVE MAKING THE ENTIRE INPUT SINE WAVE POSITIVE THROUGHOUT THE CYCLE.

Adding DC offset

Positive clamping.

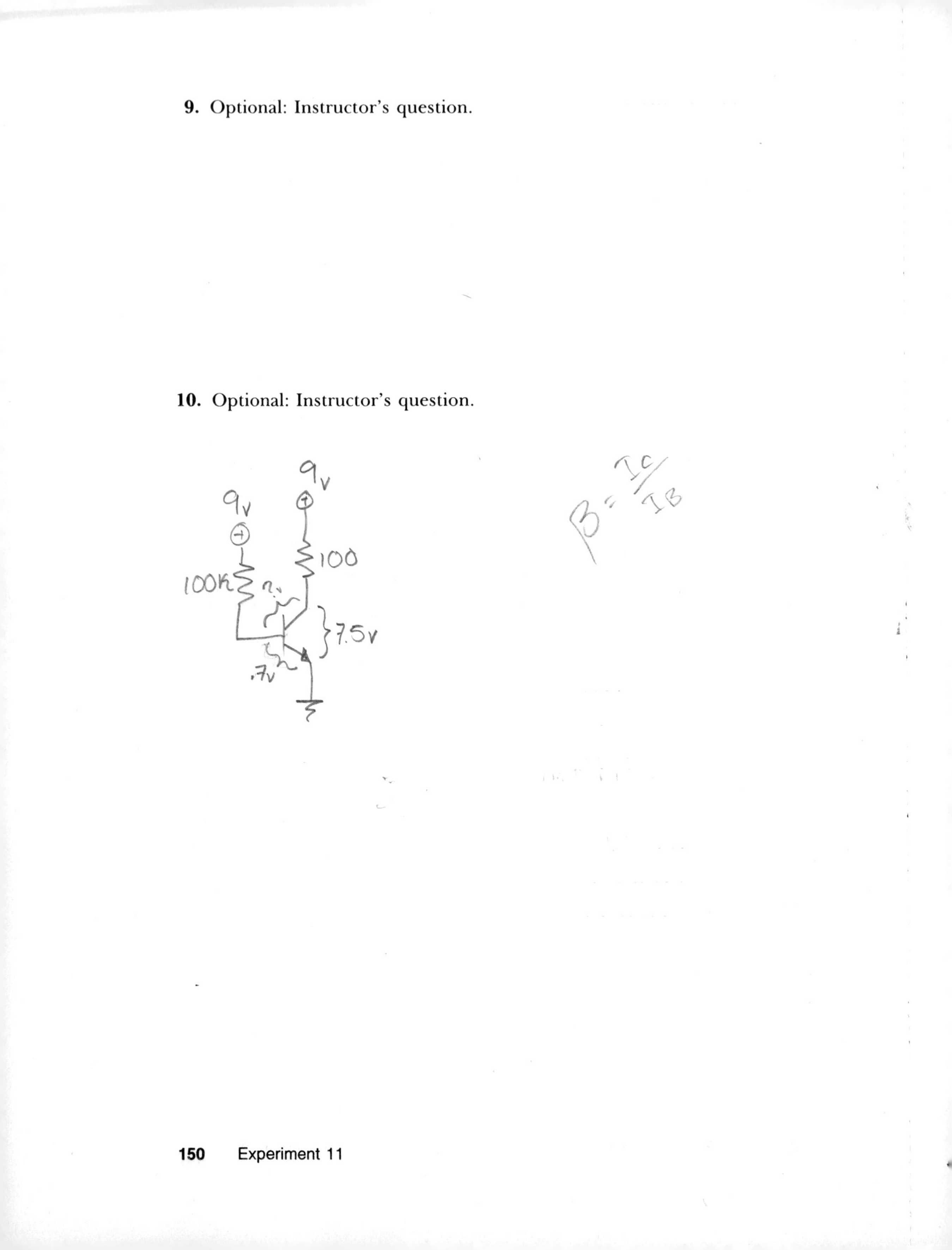

9. Optional: Instructor's question.

10. Optional: Instructor's question.

Name: TED LAIL Date: 1-28-82

QUESTIONS FOR EXPERIMENT 12

1. The transistors are silicon because:
   (a) I_C is much greater than I_B; *(b)* the collector diode is reverse-biased; *(c)* V_{BE} is close to 0.7 V; *(d)* α_{dc} is almost unity. (c)
2. How was each transistor biased?
   (a) FF; *(b)* FR; *(c)* RF; *(d)* RR. (b)
3. The collector diode was reverse-biased because:
   (a) I_C was greater than I_B; *(b)* V_{BE} was about 0.7 V; *(c)* β_{dc} was greater than unity; *(d)* V_{CB} was greater than zero. (c) d
4. In Fig. 12, the voltage across the 100-Ω resistor must equal:
   (a) $100I_C$; *(b)* 0.7 V; *(c)* $9 - 100I_C$; *(d)* $9 - V_{BE}$. (a)
5. This experiment proves the collector current of an FR-biased transistor is:
   (a) about the same as the base current; *(b)* much smaller than the emitter current; *(c)* much smaller than the base current; *(d)* much larger than the base current. (d)
6. With all transistors used, the base-emitter voltage was closest to:
   (a) 0.3 V; *(b)* 0.7 V; *(c)* 0.9 V; *(d)* 1 V. (b)
7. Write the values of Table 1 for the first transistor:

   V_{BE} .70v

   V_{CE} 6.75v

   I_C 22.2mA

   I_B .084mA

8. Write the values of Table 2 here for the first transistor:

   V_{CB} 8.59v

   I_E 22.284mA

   α_{dc} .996

   β_{dc} 264

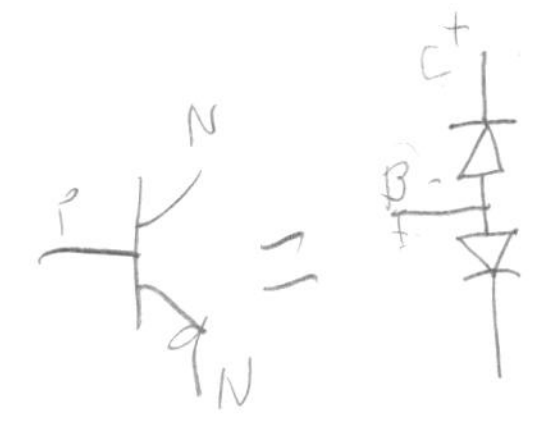

9. Optional: Instructor's question.

10. Optional: Instructor's question.

Name:  TED LAIL **Date:** 2-2-82

QUESTIONS FOR EXPERIMENT 13

1. When R in Fig. 13*a* equals 1 MΩ, base current is approximately: *(a)* 2 μA; *(b)* 4 μA; *(c)* 8 μA; *(d)* 16 μA. (c)
2. If the voltage across the 100-Ω resistor of Fig. 13*a* is 100 mV when R equals 1 MΩ, β_{dc} is close to: *(a)* 40; *(b)* 80; *(c)* 120; *(d)* 160. (c)
3. Collector curves appear upside-down in this experiment because: *(a)* the transistor is *npn*; *(b)* I_C flows up through the 100-Ω resistor; *(c)* the negative end of V_{CC} is connected to vertical input; *(d)* the voltage across the 100-Ω resistor is positive. (b)
4. The I_C values of Table 1 prove: *(a)* collector current is constant; *(b)* the transistor breaks down; *(c)* base current is constant; *(d)* I_C is almost constant when V_{CE} is between 2 and 10 V. (d)
5. Because of the values in Table 1, $V_{CE(sat)}$ must be: *(a)* less than zero; *(b)* less than 2 V; *(c)* more than 2 V; *(d)* more than 10 V. (b)
6. Table 1 proves collector breakdown voltage is: *(a)* less than zero; *(b)* less than 2 V; *(c)* more than 2 V; *(d)* more than 10 V. (d)
7. Write the Table 1 values for $R = 220$ kΩ here:

V_{CE}	2	10
I_C	8.8mA	14.4mA

8. For the two cases of the preceding question, calculate I_B and β_{dc} in Fig. 13*a*. Write the answers here:

V_{CE}	2	10
β_{dc}	340	556

9. Optional: Instructor's question.

10. Optional: Instructor's question.

Name: TED LAIL Date: 2-2-82

QUESTIONS FOR EXPERIMENT 14

1. You use Eq. (14–2) instead of Eq.(14–1) when:
   (a) accuracy is not important; *(b)* V_{BB} is extremely large; *(c)* V_{BE} makes a significant difference; *(d)* β_{dc} is small. (C)
2. When R_E increases in a base-biased circuit, which of these decreases?
   (a) I_E; *(b)* V_{BB}; *(c)* V_C; *(d)* V_{CE}. (a)
3. When R_B decreases in a base-biased circuit, which of these decreases?
   (a) V_E; *(b)* V_B; *(c)* I_E; *(d)* V_{CE}. (d)
4. When V_{BB} increases in a base-biased circuit, which of these decreases?
   (a) I_B; *(b)* V_{CE}; *(c)* I_E; *(d)* V_E. (b)
5. When β_{dc} increases in a base-biased circuit, which of these decreases?
   ~~*(a)* V_B~~; *(b)* V_E; *(c)* V_C; ~~*(d)* I_E~~. (b)() c
6. A poor way to vary the value of I_E in Fig. 14*b* is by making which of these variable?
   (a) R_B; *(b)* R_E; *(c)* R_C. (c)
7. In Table 3, V_B is approximately:
   (a) 0.3 V greater than V_E; *(b)* 0.3 V less than V_E; *(c)* 0.7 V greater than V_E; *(d)* 0.7 V less than V_E. (C)
8. The ideal and second-approximation values in Table 2 differ by approximately:
   (a) 0.1%; *(b)* 1%; *(c)* 5%; *(d)* 10%. (C)

9. Optional: Instructor's question.

10. Optional: Instructor's question.

Name: TED LAIL Date: 2-4-82

QUESTIONS FOR EXPERIMENT 15

1. The value of I_E in Fig. 15*a* is closest to:
   (a) 1 mA; *(b)* 2 mA; *(c)* 4 mA; *(d)* 7 mA. (c)
2. In Fig. 15*a*, the transistor is not saturated because:
   (a) V_{BE} is approximately 0.7 V; *(b)* voltage-divider bias can't saturate the transistor; *(c)* V_C is greater than zero; *(d)* V_{CE} is greater than 1 V. (d)
3. If β_{dc} is 100 in Fig. 15*a* and you use Eq. (7–17), the second term in the denominator adds how much?
   (a) 1.5 Ω; *(b)* 15 Ω; *(c)* 150 Ω. (b)
4. If β_{dc} equals 100, Eq. (7–17) gives an I_E in Fig. 15*a* of approximately:
   (a) 3.8 mA; *(b)* 4 mA; *(c)* 4.7 mA; *(d)* 5 mA. (b)
5. Show the following values from Tables 1 and 2:

   I_E 4.08 mA from Table 1 (calculated)

   I_E 4.06 mA from Table 2 (measured)

6. In Fig. 15*b*, the value of emitter current is closest to:
   (a) 7 mA; *(b)* 9 mA; *(c)* 9.5 mA; *(d)* 9.7 mA. (b)
7. If β_{dc} is 100 and you use Eq. (7–30), the emitter current of Fig. 15*b* is closest to:
   (a) 8.1 mA; *(b)* 8.3 mA; *(c)* 9 mA; *(d)* 10 mA. (a)
8. The voltage across R_B in Fig. 15*b* is small because:
   (a) V_B is large; *(b)* emitter current is almost constant; *(c)* V_{BE} is only 0.7 V; *(d)* $I_B R_B$ is small. (d)

9. Optional: Instructor's question.

10. Optional: Instructor's question.

Name: TED LAIL **Date:** 2-12-82

QUESTIONS FOR EXPERIMENT 16

1. When β_{dc} decreases, the R_B needed for midpoint bias in a collector-feedback circuit:
   (a) decreases; *(b)* stays the same; *(c)* increases. (c)
2. In Table 1, the I_E corresponding to a β_{dc} of 100 is closest to:
   (a) 2.5 mA; *(b)* 5 mA; *(c)* 7 mA; *(d)* 10 mA. (c)
3. In Table 1, the V_C for a β_{dc} of 25 is closest to:
   (a) 3 V; *(b)* 6 V; *(c)* 9 V; *(d)* 12 V. (d)
4. In Table 1, the V_C for a β_{dc} of 400 is closest to:
   (a) 3 V; *(b)* 6 V; *(c)* 9 V; *(d)* 12 V. (a)
5. Even though β_{dc} is between 25 and 400, Table 1 shows the transistor cannot saturate or cut off because:
   (a) V_{BE} is always close to 0.7 V; *(b)* V_{CE} is greater than 1 V and I_E is greater than zero; *(c)* the collector diode is always forward biased; *(d)* base current is never zero. (b)
6. Suppose increasing temperature causes β_{dc} to increase. Which of these does not happen with collector-feedback bias?
   (a) I_B decreases; *(b)* I_C increases; *(c)* V_C decreases; *(d)* I_B increases. (d)
7. If β_{dc} can vary from 40 to 125, the geometric average is approximately:
   (a) 60; *(b)* 70; *(c)* 80; *(d)* 90. (b)
8. If β_{dc} equals 125 and R_C is 2 kΩ, the R_B needed for midpoint bias is approximately:
   (a) 62 kΩ; *(b)* 125 kΩ; *(c)* 25 kΩ; *(d)* 250 kΩ. (d)

9. Optional: Instructor's question.

10. Optional: Instructor's question.

Name: TED LAIL Date: 2-12-82

QUESTIONS FOR EXPERIMENT 17

1. In Fig. 17*a*, V_E is:
   (a) less positive than V_C; *(b)* 0.7 V greater than V_B; *(c)* negative with respect to V_B. (b)
2. The voltage across the emitter resistor in Fig. 17*a* is 0.7 V:
   (a) less than across the 4.7-kΩ resistor; *(b)* more than across the 4.7-kΩ resistor; *(c)* less than across the collector resistor; *(d)* less than across the 2.2-kΩ resistor. (d)
3. The value of I_E in Fig. 17*a* is closest to:
   (a) 1 mA; *(b)* 2 mA; *(c)* 4 mA; *(d)* 7 mA. (C)
4. The transistor of Fig. 17*a* is not saturated because:
   (a) V_{CE} is greater than 1 V; *(b)* V_{BE} is about 0.7 V; *(c)* I_E does not equal zero. (a)
5. Write the following values of Tables 1 and 2:

   I_E 4.1 mA from Table 1 (calculated)

   I_E 4.15 mA from Table 2 (measured)

6. In Table 3, V_C is closest to:
   (a) 3 V; *(b)* 5 V; *(c)* 7 V; *(d)* 9 V. (C)
7. A 2N5226 has a minimum β_{dc} of 30 and a maximum β_{dc} of 600. To set up midpoint bias in Fig. 17*b*, R_B should have which of these standard values? *(a)* 100 kΩ; *(b)* 120 kΩ; *(c)* 130 kΩ; *(d)* 150 kΩ. (C)
8. If the 2N5226 of the preceding question is used in Fig. 17*a*, what effect does the large β_{dc} spread have?
   (a) Almost no effect on emitter current; *(b)* a significant effect on I_E; *(c)* a large effect on V_{BE}. (a)

10.9 mA 4.6% ERR.
10.4 mA

9. Optional: Instructor's question.

10. Optional: Instructor's question.

Name: TED LAIL Date: 2-17-82

QUESTIONS FOR EXPERIMENT 18

T=RC

1. A coupling capacitor ideally looks like a dc:
   (a) open and ac open; *(b)* open and ac short; *(c)* short and ac open; *(d)* short and ac short. (b)
2. A small Thevenin resistance means the bypass capacitor must be:
   (a) small; *(b)* large; *(c)* unaffected. (b)
3. The value of f_{low} in Table 1 is closest to:
   (a) 100 Hz; *(b)* 500 Hz; *(c)* 1 kHz; *(d)* 5 kHz. (c)
4. The value of f_{low} in Table 2 is closest to:
   (a) 100 Hz; *(b)* 500 Hz; *(c)* 1 kHz; *(d)* 5 kHz. (d)
5. If we want an f_{low} of 25 Hz in Fig. 18*a*, we have to change the capacitor to approximately:
   (a) 1 μF; *(b)* 2 μF; *(c)* 5 μF; *(d)* 10μF. (a)
6. In Table 1, the output voltage at f_{low} is closest to:
   (a) 0.8 V; *(b)* 0.9 V; *(c)* 0.99 V; *(d)* 1 V. (c)
7. In Table 2, the output voltage at f_{low} is closest to:
   (a) 0.01 V; *(b)* 0.15 V; *(c)* 0.99 V; *(d)* 1 V. (b)
8. Which of these is false in Fig. 18*b* at f_{low}?
   (a) Capacitor voltage is out of phase with source voltage; *(b)* total impedance seen by source is approximately 22 kΩ; *(c)* current is down about 1%; *(d)* capacitor voltage is down about 1%. (d)

T=RC
.04= 40k(c)

9. Optional: Instructor's question.

10. Optional: Instructor's question.

Name: TED LAIL Date: 2-22-82

SKS

QUESTIONS FOR EXPERIMENT 19

1. In Table 1, each time emitter current is cut in half, ac emitter resistance:
   (a) goes down by a factor of two; *(b)* doubles; *(c)* quadruples; *(d)* stays the same. (b)
2. In Fig. 19*a*, the positive end of the $-V_{EE}$ supply is:
   (a) grounded; *(b)* connected to the 10-kΩ resistor; *(c)* connected to the emitter. (a)
3. When $-V_{EE}$ equals -10.7 V in Fig. 19*b*, V_{CE} equals:
   (a) 0.7 V; *(b)* 1 V; *(c)* 5 V; *(d)* 8 V. (d)
4. The reason the biasing 10-kΩ resistor of Fig. 19*a* is unimportant in Fig. 19*c* is:
   (a) V_{BE} is about 0.7 V; *(b)* negligible dc current goes through the coupling capacitor; *(c)* most of the 0.1 mA goes through r_e'; *(d)* r_e' is not in parallel with this resistor. (c)
5. Because r_e' is only a small part of the resistance in Fig. 19*c*, the circuit driving r_e' is approximately an ideal:
   (a) voltage source; *(b)* current source; *(c)* resistance. (b)
6. According to Tables 1 and 2, the value of v_{be} for an I_E of 0.5 mA is closest to:
   (a) 2.5 mV; *(b)* 5 mV; *(c)* 50 mV; *(d)* 250 mV. (b)
7. Table 2 confirms that r_e' is:
   (a) 25 Ω; *(b)* equal to v_{be}/i_b; *(c)* inversely proportional to v_{be}; *(d)* inversely proportional to I_E. (d)
8. Which of the following cannot account for the different r_e' values in Tables 1 and 2?
   (a) Eq. (8–5) is an ideal formula; *(b)* tolerance of resistors; *(c)* the base-emitter junction is not rectangular; *(d)* transistor temperature is near 100°C. (d)

9. Optional: Instructor's question.

10. Optional: Instructor's question.

Name: TED LAIL Date: 2-23-82

QUESTIONS FOR EXPERIMENT 20

1. In Fig. 20, the voltage gain from base to collector is ideally around: *(a)* 10; *(b)* 50; *(c)* 100; *(d)* 150. ( )
2. The reason point B has an ac signal of 20 mV pp is because the input coupling capacitor looks like:
(a) a resistor; *(b)* a dc short; *(c)* a dc open; *(d)* an ac short. ( )
3. The reason point I has no dc voltage is because the input coupling capacitor looks like:
(a) a resistor; *(b)* a dc short; *(c)* a dc open; *(d)* an ac short. ( )
4. Ideally, all the ac input signal appears across the base-emitter diode in Fig. 20 because:
(a) the emitter is ac grounded; *(b)* point E is at dc ground; *(c)* there's no ac current through the bypass capacitor; *(d)* there's ac current through the 10-kΩ emitter resistor. ( )
5. The value of A_{calc} in Table 2 is closest to:
(a) 10; *(b)* 20; *(c)* 50; *(d)* 100. ( )
6. Which of the following is not a possible reason A_{calc} differs from A in Table 2?
(a) r'_e is not exactly equal to (25 mV)/I_E; *(b)* tolerance of resistors; *(c)* r'_b was neglected; *(d)* V_{BE} is 0.7 V. ( )
7. If the bypass capacitor opens in Fig. 20, the voltage gain from base to collector is approximately:
(a) 0.1; *(b)* 0.2; *(c)* 1; *(d)* 5. ( )
8. If the final 4.7-kΩ resistor of Fig. 20 opens, the voltage gain from base to collector increases to about:
(a) 250; *(b)* 180; *(c)* 120; *(d)* 80. ( )

9. Optional: Instructor's question.

10. Optional: Instructor's question.

Name: TED LAIL Date: 2-25-82

5KS

QUESTIONS FOR EXPERIMENT 21

1. In Fig. 21*a*, V_C is closest to:
   (a) 2 V; *(b)* 5 V; *(c)* 7 V; *(d)* 8 V. (b)
2. In Fig. 21*a*, V_E is closest to:
   (a) −0.7 V; *(b)* 0.7 V; *(c)* −2 V; *(d)* 2 V. (a)
3. The ac resistance seen by the collector in Fig. 21*a* is approximately:
   (a) 1.5 kΩ; *(b)* 1.8 kΩ; *(c)* 2.2 kΩ; *(d)* 4.7 kΩ. (a)
4. The value of $z_{in(base)}$ in Table 1 for a β of 100 is approximately:
   (a) 30 Ω; *(b)* 100 Ω; *(c)* 3 kΩ; *(d)* 8.2 kΩ. (c)
5. When the decade resistance box R equals the value of $z_{in(base)}$, how much voltage is across R?
   (a) 5 mV; *(b)* 10 mV; *(c)* 15 mV; *(d)* 20 mV. (a)
6. If β is 100 and the bypass capacitor of Fig. 21 opens, $z_{in(base)}$ will be approximately:
   (a) 2.5 kΩ; *(b)* 17.5 kΩ; *(c)* 20 kΩ; *(d)* 1 MΩ. (d)
7. If β is 100 and the 2.2-kΩ resistor of Fig. 21*a* opens, the value of $z_{in(base)}$ is closest to:
   (a) 25 Ω; *(b)* 3 kΩ; *(c)* 17.5 kΩ; *(d)* 20 kΩ. (b)

OK

8. Write the values of $z_{in(base)}$ from Table 2:

   $z_{in(base)}$ 3.44K, 6.8K, 12.9K?

   β = 360

9. Optional: Instructor's question.

10. Optional: Instructor's question.

Name: TED LAIL Date: 3-1-82

S/S

QUESTIONS FOR EXPERIMENT 22

1. For the voltage of an emitter follower to equal unity, which of these must be zero?
   (a) r'_e; *(b)* r_E; *(c)* β; *(d)* I_E. (a)
2. The ac resistance seen by the emitter in Fig. 22*a* is closest to:
   (a) 20 Ω; *(b)* 90 Ω; *(c)* 110 Ω; *(d)* 1 kΩ. (b)
3. The value of $z_{\text{in(base)}}$ for a β of 100 in Table 1 is closest to:
   (a) 5 kΩ; *(b)* 9 kΩ; *(c)* 15 kΩ; *(d)* 50 kΩ. (b)
4. The value of A in Table 1 is approximately:
   (a) 0.5; *(b)* 0.65; *(c)* 0.8; *(d)* 1. (c)
5. If β is 100 in Fig. 22*a*, the power gain from base to emitter is around:
   (a) 50; *(b)* 80; *(c)* 100; *(d)* 125. (b)
6. Ideally, the collector in Fig. 22*b* sees an ac resistance of:
   (a) zero; *(b)* 10 kΩ; *(c)* 10 kΩ in parallel with 1 MΩ. (a)
7. If the *npn* and *pnp* transistors have identical characteristics except for oppositely doped regions, then which of these differs in Figs. 22*a* and *b*?
   (a) $z_{\text{in(base)}}$; *(b)* ac emitter voltage; *(c)* A; *(d)* dc collector-to ground voltage. (a) d
8. If a Darlington pair with a β of 5000 is used in Fig. 22*a*, $z_{\text{in(base)}}$ increases to roughly:
   (a) 50 kΩ; *(b)* 250 kΩ; *(c)* 500 kΩ; *(d)* 1 MΩ. (c)

G = B (A) =

9. Optional: Instructor's question.

10. Optional: Instructor's question.

Name: TED LAIL Date: 2-5-82

QUESTIONS FOR EXPERIMENT 23

1. The dc collector-to-ground voltage in Fig. 23*a* is closest to:
   (a) 2 V; *(b)* 5 V; *(c)* 7 V; *(d)* 9 V. (c)
2. The $z_{in(emitter)}$ of Table 1 is nearest to:
   (a) 10 Ω; *(b)* 12 Ω; *(c)* 14 Ω; *(d)* 16 Ω. (c)
3. The voltage gain *A* of Table 1 is approximately:
   (a) 27; *(b)* 50; *(c)* 73; *(d)* 94. (b)
4. If we change the 4.7-kΩ resistor of Fig. 23*a* to 9.1 kΩ, the voltage gain approximately:
   (a) decreases by a factor of two; *(b)* doubles; *(c)* triples. (a)
5. If the output coupling capacitor shorts, the voltage gain *A* of Fig. 23*a* equals approximately:
   (a) 27; *(b)* 50; *(c)* 73; *(d)* 94. (b)
6. Visualize the ac equivalent circuit of Fig. 23*a*. The dc sources appear as:
   (a) 9-V sources; *(b)* open circuits; *(c)* resistors; *(d)* short circuits. (d)
7. In Fig. 23*a*, the voltage gain from point *A* to the collector is closest to:
   (a) 0.7; *(b)* 1; *(c)* 2; *(d)* 10. (a)
8. In Fig. 23*b*, the ac voltage across r_e' should be approximately:
   (a) 1 mV; *(b)* 10 mV; *(c)* 14 mV; *(d)* 20 mV. (c)

9. Optional: Instructor's question.

10. Optional: Instructor's question.

Name: TED LAIL Date: 3-11-82

SKS

QUESTIONS FOR EXPERIMENT 24

1. In Fig. 24*a,* the current through the 100-Ω resistor is: *(a)* dc current; *(b)* ac current; *(c)* total current. (C)
2. The reason the ac load line appears upside-down in this experiment is because the: *(a)* transistor is *npn;* *(b)* collector current flows up through the 100-Ω resistor; *(c)* negative end of the 9-V supply is connected to the vertical input; *(d)* voltage across the 100-Ω resistor is positive. (b)
3. When the audio generator is turned down to zero, the current through the 100-Ω resistor is identical to the: *(a)* ac collector current; *(b)* current through the 10-kΩ resistor; *(c)* current through the decade resistance; *(d)* base current. (b)
4. The voltage to the vertical input equals the: *(a)* current through the 100-Ω resistor; *(b)* collector voltage; *(c)* ac collector current; *(d)* product of total collector current and 100 Ω. (d)
5. The voltage to the horizontal input equals the: *(a)* base-emitter voltage; *(b)* collector-emitter voltage; *(c)* collector-base voltage; *(d)* vertical-input voltage. (b)
6. As the ac load resistance increases, the ac load line becomes: *(a)* straighter; *(b)* more vertical; *(c)* more horizontal. (C)
7. If $I_{CQ} = 0.6$ mA and $V_{CEQ} = 3$ V in Fig. 24*a,* the ac collector resistance that results in a centered *Q* point is: *(a)* 2 kΩ; *(b)* 2.5 kΩ; *(c)* 5 kΩ; *(d)* 10 kΩ. (C)
8. Write the values of Table 2 here:

R	r_C	$I_{C(sat)}$	$V_{CE(cutoff)}$
2.5 kΩ	2K	2.75 mA	5.4
10 kΩ	5K	1.3 mA	6.8
1 MΩ	9.9K	.9 mA	8.7

9. Optional: Instructor's question.

10. Optional: Instructor's question.

Name: TED LAIL Date: 3-10-82

QUESTIONS FOR EXPERIMENT 25

1. The quiescent collector current in Fig. 25 is approximately:
(a) 5 mA; *(b)* 10 mA; *(c)* 20 mA; *(d)* 25 mA. (C)
2. The value of V_{CEQ} in Fig. 25 is closest to:
(a) 5 V; *(b)* 10 V; *(c)* 15 V; *(d)* 20 V. (a)
3. The value of r_E that centers the Q point in Fig. 25 is closest to:
(a) 100 Ω; *(b)* 250 Ω; *(c)* 350 Ω; *(d)* 470 Ω. (b)
4. The quiescent dissipation of the transistor in Fig. 25 is nearest to:
(a) 50 mW; *(b)* 100 mW; *(c)* 250 mW; *(d)* 500 mW. (b)
5. When R is 100 Ω in Fig. 25, we get cutoff clipping first. The textbook illustrates this case in Fig.:
(a) 10-8*a*; *(b)* 10-8*b*; *(c)* 10-8*c*; *(d)* 10-8*d*. (a)
6. When R is infinity, we get saturation clipping first. The textbook illustrates this in Fig.:
(a) 10-8*a*; *(b)* 10-8*b*; *(c)* 10-8*c*; *(d)* 10-8*d*. (b)
7. Table 3 indicates the peak-to-peak voltage is greater than the rms voltage by roughly a factor of:
(a) 1; *(b)* 2; *(c)* 3; *(d)* 4. (C)
8. The maximum load power in Table 3 is closest to:
(a) 10 mW; *(b)* 15 mW; *(c)* 30 mW; *(d)* 50 mW. (d)

9. Optional: Instructor's question.

10. Optional: Instructor's question.

Name: ____________________ Date: __________

QUESTIONS FOR EXPERIMENT 26

1. The $I_{C(sat)}$ of Table 1 is approximately:
   (a) 50 mA; *(b)* 100 mA; *(c)* 150 mA; *(d)* 200 mA. ( )
2. The distortion observed with V_{CC} equal to 5 V is normally referred to as:
   (a) nonlinear distortion; *(b)* crossover distortion; *(c)* harmonic distortion; *(d)* frequency distortion. ( )
3. The V_{BE} of Table 2 is closest to:
   (a) 0.5 V; *(b)* 0.55 V; *(c)* 0.65 V; *(d)* 0.7 V. ( )
4. If V_{CC} is 10 V in Fig. 26*a*, V_{CEQ} is:
   (a) 5 V; *(b)* 10 V; *(c)* 15 V; *(d)* 20 V. ( )
5. Theoretically, the maximum peak-to-peak output voltage for a V_{CC} of 10 V in Fig. 26*a* is:
   (a) 5 V; *(b)* 10 V; *(c)* 15 V; *(d)* 20 V. ( )
6. Table 2 shows the peak-to-peak voltage is greater than the rms voltage by roughly a factor of:
   (a) 1; *(b)* 2; *(c)* 3; *(d)* 4. ( )
7. The load power in Table 2 is closest to:
   (a) 10 mW; *(b)* 40 mW; *(c)* 70 mW; *(d)* 100 mW. ( )
8. In this experiment, I_{CQ} was set at 1 mA. This represents a trickle bias of what percent of $I_{C(sat)}$?
   (a) 1%; *(b)* 2%; *(c)* 4%; *(d)* 5%. ( )

9. Optional: Instructor's question.

10. Optional: Instructor's question.

Name: ______________________________ **Date:** ____________

QUESTIONS FOR EXPERIMENT 27

1. The $I_{C(sat)}$ in Fig. 27*a* is closest to:
   (a) 20 mA; *(b)* 30 mA; *(c)* 50 mA; *(d)* 100 mA. ( )
2. The V_{CEQ} in Fig. 27*a* is:
   (a) 0; *(b)* 5 V; *(c)* 10 V; *(d)* 15 V. ( )
3. The period of the pulses in Table 1 is:
   (a) 10 μs; *(b)* 100 μs; *(c)* 1 ms; *(d)* 10 ms. ( )
4. The duty cycle in Table 1 is closest to:
   (a) 1%; *(b)* 15%; *(c)* 50%; *(d)* 100%. ( )
5. The base signal is negatively clamped because the:
   (a) capacitor charges; *(b)* transistor turns on at the negative input peak; *(c)* collector diode is part of the clamper; *(d)* 100-kΩ resistor has a positive dc voltage across it. ( )
6. In Fig. 27*b*, each milliampere of collector current produces how much vertical input voltage?
   (a) 100 mV; *(b)* 500 mV; *(c)* 1 V; *(d)* 5 V. ( )
7. The $I_{C(sat)}$ in Table 2 is closest to:
   (a) 1 mA; *(b)* 2 mA; *(c)* 4 mA; *(d)* 8 mA. ( )
8. If the signal from the audio generator is reduced to zero in Fig. 27*b*, the oscilloscope will display a:
   (a) load line; *(b)* dot at zero; *(c)* dot at $I_{C(sat)}$; *(d)* dot at $V_{CE(cutoff)}$. ( )

9. Optional: Instructor's question.

10. Optional: Instructor's question.

Name: ______________________ Date: __________

QUESTIONS FOR EXPERIMENT 28

1. In Fig. 28*a*, each milliampere of drain current produces a vertical input of:
   (a) 1 mV; *(b)* 10 mV; *(c)* 100 mV; *(d)* 1 V. ( )
2. If the I_{DSS} of the JFET is 16 mA, the vertical input in Fig. 28*a* for a V_{DS} greater than pinchoff will be:
   (a) 16 μV; *(b)* 16 mV; *(c)* 1.6 V; *(d)* 16 V. ( )
3. The drain curve appeared upside-down in this experiment because:
   (a) an *n*-channel JFET was used; *(b)* I_D flows up through the 100-Ω resistor; *(c)* the negative end of V_{DD} is connected to the vertical input; *(d)* the voltage across the drain-source terminals is positive. ( )
4. The pinchoff voltage was in the vicinity of:
   (a) 0 to 1 V; *(b)* 2 to 6 V; *(c)* 5 to 8 V; *(d)* 8 to 10 V. ( )
5. The pinchoff voltage theoretically equals:
   (a) 2 V; *(b)* $V_{GS(\text{off})}$; *(c)* $V_{GS(\text{off})}/4$; *(d)* V_{DS}. ( )
6. In Fig. 28*d*, the input audio signal is:
   (a) negatively clamped; *(b)* positively clamped; *(c)* negatively clipped; *(d)* negatively peak-rectified. ( )
7. The voltage applied to the horizontal input in Fig. 28*d* is:
   (a) V_{GS}; *(b)* V_{DS}; *(c)* $100I_D$; *(d)* none of these. ( )
8. If the horizontal input is disconnected in Fig. 28*d* and the audio generator turned down to zero, you will see a:
   (a) vertical line; *(b)* horizontal line; *(c)* spot at zero; *(d)* vertically deflected spot. ( )

9. Optional: Instructor's question.

10. Optional: Instructor's question.

Name: ______________________ Date: __________

QUESTIONS FOR EXPERIMENT 29

1. Step 1 and the V_S value for voltage-divider bias in Table 1 prove that gate-source diode is:
   (a) reverse-biased; *(b)* forward-biased; *(c)* zero-biased. ( )
2. In Fig. 29*a*, the current through the source resistor produces a source voltage that is:
   (a) negative; *(b)* positive but less than V_G; *(c)* positive but greater than V_G; *(d)* more negative than gate voltage. ( )
3. In the self-biased circuit of Fig. 29*b*, the gate voltage to ground is:
   (a) approximately zero; *(b)* about 1 V; *(c)* equal to the source voltage; *(d)* negative. ( )
4. Current through the source resistor results in a gate voltage that is:
   (a) positive with respect to the source; *(b)* negative with respect to the source; *(c)* equal to the source. ( )
5. The drain current of Fig. 29*c* is closest to:
   (a) 0; *(b)* 1 mA; *(c)* 2 mA; *(d)* 4 mA. ( )
6. The drain-to-ground voltage of Fig. 29*c* is nearest to:
   (a) 0; *(b)* 5 V; *(c)* 10 V; *(d)* 15 V. ( )
7. For the current-source bias of Fig. 29*d*, V_{DS} is closest to:
   (a) 5 V; *(b)* 10 V; *(c)* 15 V; *(d)* 20 V. ( )
8. Table 1 proves the *npn* transistor of Fig. 29*d* is not saturated because:
   (a) V_{DS} is greater than the pinchoff voltage; *(b)* $V_{GS(\text{off})}$ is greater than -2 V; *(c)* V_S is greater than the $V_{CE(\text{sat})}$ of typical transistors; *(d)* V_D is well above the $V_{CE(\text{sat})}$ of any normal transistor. ( )

9. Optional: Instructor's question.

10. Optional: Instructor's question.

Name: ______________________ Date: __________

QUESTIONS FOR EXPERIMENT 30

1. No input coupling capacitor is needed in Fig. 30*a* or *b* because: *(a)* the gate is at ac ground; *(b)* the gate is at dc ground; *(c)* the source is bypassed; *(d)* there's no output coupling capacitor. ( )
2. The input resistance looking into the JFET amplifier or source follower of Fig. 30 is: *(a)* 2.2 kΩ; *(b)* 4.7 kΩ; *(c)* 220 kΩ; *(d)* infinity. ( )
3. When the bypass capacitor is opened in Fig. 30*a*, the voltage gain drops to approximately: *(a)* 1; *(b)* 2; *(c)* 3; *(d)* 4. ( )
4. The minimum voltage gain for A_1 in Table 1 is approximately: *(a)* 7; *(b)* 10; *(c)* 24; *(d)* 30. ( )
5. The minimum voltage gain for A_2 in Table 1 is approximately: *(a)* 0.333; *(b)* 0.5; *(c)* 0.75; *(d)* 1. ( )
6. The value of r_S in Fig. 30*b* is nearest to: *(a)* 680 Ω; *(b)* 1 kΩ; *(c)* 2.2 kΩ; *(d)* 220 kΩ. ( )
7. If g_m equals 5000 μS, the voltage gain of the source follower (Fig. 30*b*) is closest to: *(a)* 0.475; *(b)* 0.625; *(c)* 0.775; *(d)* 1. ( )
8. Write the following values of Table 2 here:

   g_m ______________________

   v_{out} (no bypass) ______________________

9. Optional: Instructor's question.

10. Optional: Instructor's question.

Name: ______________________________ **Date:** ____________

QUESTIONS FOR EXPERIMENT 31

1. Which of the following is closest to the bel voltage gain of a lightly loaded emitter follower?
   (a) 0 dB; *(b)* −3 dB; *(c)* −6 dB; *(d)* −10 dB. ( )
2. Negative values of bel voltage gain always mean the ordinary voltage gain is:
   (a) negative; *(b)* phase inverted; *(c)* less than unity; *(d)* less than zero. ( )
3. The value of A' for an R of 100 kΩ in Table 1 is:
   (a) −3 dB; *(b)* −6 dB; *(c)* −10 dB; *(d)* −20 dB. ( )
4. The value of A' for an R of 11.1 kΩ in Table 1 is:
   (a) −3 dB; *(b)* −6 dB; *(c)* −10 dB; *(d)* −20 dB. ( )
5. The ordinary voltage gain of each voltage divider in Fig. 31*b* is:
   (a) −6 dB; *(b)* −10 dB; *(c)* −18 dB; *(d)* 0.5. ( )
6. The total bel voltage gain in Fig. 31*b* is:
   (a) −6 dB; *(b)* −12 dB; *(c)* −18 dB; *(d)* 0.125. ( )
7. The total ordinary voltage gain in Fig. 31*b* is:
   (a) 0.125; *(b)* 0.25; *(c)* 0.5; *(d)* 0.75. ( )
8. In Table 3, each decibel reading differs by how much from the preceding reading?
   (a) −3 dB; *(b)* −6 dB; *(c)* −9 dB; *(d)* −18 dB. ( )

9. Optional: Instructor's question.

10. Optional: Instructor's question.

Name: ______________________ Date: ____________

QUESTIONS FOR EXPERIMENT 32

1. The input Miller resistance in Table 1 is:
   (a) 10 Ω; *(b)* 1 kΩ; *(c)* 10 kΩ; *(d)* 100 kΩ. ( )
2. The input Miller capacitance in Table 1 is:
   (a) 0.001 μF; *(b)* 0.01 μF; *(c)* 0.1 μF; *(d)* 1 μF. ( )
3. In Fig. 32*a,* the 10-MΩ biasing resistor produces an input Miller resistance of:
   (a) 100 Ω; *(b)* 10 kΩ; *(c)* 1 MΩ; *(d)* 10 MΩ. ( )
4. The input resistance R_{in} of Fig. 32*b* is approximately:
   (a) 10 kΩ; *(b)* 100 kΩ; *(c)* 1 MΩ; *(d)* 10 MΩ. ( )
5. The input capacitance C_{in} of Fig. 32*c* is approximately:
   (a) 0.001 μF; *(b)* 0.01 μF; *(c)* 0.1 μF; *(d)* 1 μF. ( )
6. In Fig. 32*b,* v_A is approximately two times v_{in} because the:
   (a) 10-kΩ resistor matches in input resistance; *(b)* input resistance R_{in} is 100 kΩ; *(c)* voltage gain is 9; *(d)* input Miller resistance is smaller than 100 kΩ. ( )
7. In Fig. 32*c,* v_A is approximately two times v_{in} because the:
   (a) input resistance is 10 kΩ; *(b)* input capacitance C_{in} is 0.001 μF; *(c)* voltage gain is 9; *(d)* 0.01-μF capacitor approximately matches the value of C_{in}. ( )
8. This experiment proves how much of the Miller theorem?
   (a) All of it; *(b)* the input part; *(c)* the output part. ( )

9. Optional: Instructor's question.

10. Optional: Instructor's question.

Name: ______________________ **Date:** ____________

QUESTIONS FOR EXPERIMENT 33

1. The critical frequency recorded at the top of Table 1 is:
   (a) 15.9 Hz; *(b)* 100 Hz; *(c)* 159 Hz; *(d)* 1.59 kHz. ( )
2. The internal resistance of a typical audio generator is under 1000 Ω. If this is included in the calculations for critical frequency in Fig. 33*a,* what happens?
   (a) The critical frequency is slightly lower; *(b)* the critical frequency is slightly higher; *(c)* the output level increases; *(d)* the input level increases. ( )
3. If the capacitor of Fig. 33*a* has a tolerance of ±10%, the critical frequency may be:
   (a) off as much as 20%; *(b)* as low as 1.3 kHz; *(c)* as high as 1.75 kHz. ( )
4. If the capacitor in a lag network is leaky (poor dielectric setting up a low-resistance parallel path), the critical frequency will:
   (a) decrease; *(b)* increase; *(c)* either; *(d)* neither. ( )
5. For which frequency in Table 1 is the ordinary voltage gain equal to 0.707?
   (a) $0.1f_c$; *(b)* f_c; *(c)* $10 f_c$; *(d)* $100f_c$. ( )
6. For which frequency in Table 1 is the ordinary voltage gain equal to 0.1?
   (a) $0.1f_c$; *(b)* f_c; *(c)* $10f_c$; *(d)* $100f_c$. ( )
7. The value of x in Table 2 for a frequency of f_c is approximately:
   (a) $T/8$; *(b)* $T/2$; *(c)* $2T$; *(d)* $4T$. ( )
8. The value of T in Table 2 for a frequency of $10f_c$ is nearest to:
   (a) 0.6 ms; *(b)* 6 ms; *(c)* 6 μs; *(d)* 60 μs. ( )

9. Optional: Instructor's question.

10. Optional: Instructor's question.

Name: ______________________ Date: ____________

QUESTIONS FOR EXPERIMENT 34

1. The bypass capacitor of Fig. 34 ideally affects the critical frequency of:
   (a) the gate lag network; *(b)* the drain lag network; *(c)* both networks; *(d)* neither network. ( )
2. The critical frequency in Table 1 is closest to:
   (a) 1 kHz; *(b)* 1.4 kHz; *(c)* 2.6 kHz; *(d)* 17 kHz. ( )
3. The critical frequency in Table 2 is approximately:
   (a) 1 kHz; *(b)* 1.4 kHz; *(c)* 2.6 kHz; *(d)* 17 kHz. ( )
4. With a g_m of 2000 μS, the value of $g_m r_D$ is approximately:
   (a) 5; *(b)* 8.2; *(c)* 9.4; *(d)* 10.8. ( )
5. The gate critical frequency of Table 4 may differ from that of Table 1. Which of the following is not a possible cause?
   (a) g_m may be different from 2000 μS; *(b)* capacitance of ac millivoltmeter and stray-wiring capacitances; *(c)* tolerance of R's and C's; *(d)* tolerance of bypass capacitor. ( )
6. In Fig. 34, suppose we remove the 1000-pF capacitors. Suppose further that the internal capacitances C_{gs}, C_{gd}, and C_{ds} are 10 pF each. In this case, the critical frequency of the gate lag network is closest to:
   (a) 100 kHz; *(b)* 140 kHz; *(c)* 250 kHz; *(d)* 2 MHz. ( )
7. If a square-wave input is used in Fig. 34 instead of a sine wave, the risetime of the output will be closest to:
   (a) 250 μs; *(b)* 500 μs; *(c)* 1000 μs; *(d)* 2500 μs. ( )
8. If the 4.7-kΩ resistor in Fig. 34 is increased in value, the critical frequency of which of these lag networks is affected?
   (a) Gate; *(b)* drain; *(c)* both; *(d)* neither. ( )

9. Optional: Instructor's question.

10. Optional: Instructor's question.

Name: ______________________ Date: __________

QUESTIONS FOR EXPERIMENT 35

1. The critical frequency in Table 1 is:
   (a) 723 Hz; *(b)* 1 kHz; *(c)* 7.23 kHz; *(d)* 100 kHz. ( )
2. The risetime in Table 1 is closest to:
   (a) 5 μs; *(b)* 50 μs; *(c)* 500 μs; *(d)* 5 ms. ( )
3. Disregarding measurement errors and tolerance of parts, the values of Tables 1 and 2 prove:
   (a) $f_C = 1/2\pi RC$; *(b)* $T_R = 0.35/f_C$; *(c)* both of these; *(d)* neither of these. ( )
4. If a step voltage drives an amplifier and the output has a risetime of 10 μs, what is the bandwidth of the amplifier (assuming one dominant lag network)?
   (a) 3.5 kHz; *(b)* 35 kHz; *(c)* 350 kHz; *(d)* 3.5 MHz. ( )
5. The sagtime of Table 3 is closest to:
   (a) 5 ps; *(b)* 50 ns; *(c)* 500 μs; *(d)* 5 ms. ( )
6. Table 4 proves which of these?
   (a) The ratio of sagtime to critical frequency is 0.35; *(b)* sagtime is directly proportional to critical frequency; *(c)* the ratio of critical frequency to sagtime is 0.35; *(d)* the product of sagtime and critical frequency is 0.35. ( )
7. A perfect square wave drives an amplifier with one dominant lag network and one dominant lead network. If T_S equals 17.5 ms and T_R is 17.5 μs, the bandwidth of the amplifier is between:
   (a) 2 and 2000 Hz; *(b)* 20 and 20,000 Hz; *(c)* 20 Hz and 200 kHz; *(d)* 200 Hz and 2 MHz. ( )
8. An amplifier with one dominant lag network has an upper cutoff frequency of 10 MHz. If a step voltage drives this amplifier, the output risetime will be:
   (a) 0.35 μs; *(b)* 35 ns; *(c)* 350 ns; *(d)* 3.5 μs. ( )

9. Optional: Instructor's question.

10. Optional: Instructor's question.

Name: ______________________________ **Date:** ____________

QUESTIONS FOR EXPERIMENT 36

1. The voltage gain in Table 1 is closest to:
   (a) 50; *(b)* 150; *(c)* 250; *(d)* 350. ( )
2. Voltage v_e in Table 2 is in phase with:
   (a) v_1; *(b)* v_{c2}; *(c)* the external trigger signal; *(d)* all of these. ( )
3. The only voltage out of phase with the audio-generator signal in Fig. 36*a* is:
   (a) v_{c2}; *(b)* v_{c1}; *(c)* v_1; *(d)* v_e. ( )
4. Table 3 proves v_e is approximately:
   (a) half of v_1; *(b)* in phase with v_1; *(c)* out of phase with v_1; *(d)* equal to v_1. ( )
5. Which of these trigonometric indentities corresponds to Fig. 36*b*?
   (a) $\sin\theta + \sin\theta = 2\sin\theta$; *(b)* $\sin\theta - \sin\theta = 2\sin\theta$; *(c)* $\sin\theta - (-\sin\theta) = 2\sin\theta$; *(d)* $\sin\theta - \sin\theta = -2\sin\theta$. ( )
6. If β is 100 for each transistor in Fig 36*a*, the input impedance looking into the base of Q_1 is approximately:
   (a) 2.8 kΩ; *(b)* 5.6 kΩ; *(c)* 7.5 kΩ; *(d)* 10 kΩ. ( )
7. A 2N2920 is a matched dual transistor (a pair of almost identical transistors in a single package). Suppose we use a 2N2920 in Fig. 36*a*. If the output offset voltage is 0.12 V and the voltage gain is 400, the input offset voltage is:
   (a) 3 μV; *(b)* 30 μV; *(c)* 300 μV; *(d)* 3000 μV. ( )
8. In Fig. 36*a*, suppose a common-mode input of 10 mV rms produces a v_{out} of 1 mV. If the voltage gain is 400, what is the common-mode rejection ratio in decibels?
   (a) 72 dB; *(b)* 80 dB; *(c)* 86 dB; *(d)* 92 dB. ( )

9. Optional: Instructor's question.

10. Optional: Instructor's question.

Name: ______________________________ **Date:** ____________

QUESTIONS FOR EXPERIMENT 37

1. The bypass capacitors on each supply pin prevent:
   (a) supply-voltage decrease; *(b)* oscillations; *(c)* input-signal loss; *(d)* increase in voltage. ( )
2. The maximum possible swing in output voltage in Fig. 37 is closest to:
   (a) ±1 V; *(b)* ±5 V; *(c)* ±15 V; *(d)* ±30 V. ( )
3. The voltage gain of Fig. 37 with $R = 10$ kΩ and a 741C is nearest to:
   (a) 1; *(b)* 10; *(c)* 100; *(d)* 1000. ( )
4. The voltage gain with $R = 100$ kΩ and a 741C is closest to:
   (a) 1; *(b)* 10; *(c)* 100; *(d)* 1000. ( )
5. The 10-V power bandwidth of the 741C is closest to:
   (a) 100 Hz; *(b)* 1 kHz; *(c)* 10 kHz; *(d)* 100 kHz. ( )
6. When the 318C is substituted for a 741C, which of the following changes?
   (a) Slew rate; *(b)* voltage gain; *(c)* supply voltage. ( )
7. Compared to the 741C, the 318C has a 10-V power bandwidth that is:
   (a) much smaller; *(b)* about the same; *(c)* much larger. ( )
8. One way to increase the power bandwidth is to:
   (a) increase supply voltages; *(b)* reduce output peak voltage; *(c)* increase the input frequency; *(d)* increase the input amplitude. ( )

9. Optional: Instructor's question.

10. Optional: Instructor's question.

Name: ______________________ Date: __________

QUESTIONS FOR EXPERIMENT 38

1. The approximate value of A_{SP} in Fig. 38*a* is:
   (a) 100; *(b)* 1000; *(c)* 10,000; *(d)* 100,000. ( )
2. If the 741C in Fig. 38*a* has an internal voltage gain of 100,000, the sacrifice factor is:
   (a) 100; *(b)* 1000; *(c)* 10,000; *(d)* 100,000. ( )
3. If the z_{out} of the 741C is 300 Ω, the z_{out}(SP) in Fig. 38*a* for an internal voltage gain of 100,000 is:
   (a) 0.3 Ω; *(b)* 3 Ω; *(c)* 30 Ω; *(d)* 300 Ω. ( )
4. The value of v_{in} in Table 2 is approximately:
   (a) 1 mV; *(b)* 10 mV; *(c)* 100 mV; *(d)* 1 V. ( )
5. When the 1-MΩ resistor is connected in Fig. 38*b*, the voltage to the noninverting input is:
   (a) 9 μV; *(b)* 90 μV; *(c)* 900 μV; *(d)* 9000 μV. ( )
6. In step 9, the VOM reads closest to:
   (a) 9 mV; *(b)* 90 mV; *(c)* 900 mV; *(d)* 9 V. ( )
7. If the VOM is on its 3-V range in Fig. 38*b*, you can measure dc voltages to the noninverting input from zero to:
   (a) 30 μV; *(b)* 3 mV; *(c)* 30 mV; *(d)* 3 V. ( )
8. Instead of using a VOM in Fig. 38*b*, you can use the oscilloscope to measure dc output voltage. If the vertical sensitivity is 10 mV/cm, how much dc voltage to the noninverting input produces 1 cm of deflection?
   (a) 0.1 μV; *(b)* 1 μV; *(c)* 10 μV; *(d)* 100 μV. ( )

9. Optional: Instructor's question.

10. Optional: Instructor's question.

Name: ______________________________ **Date:** ____________

QUESTIONS FOR EXPERIMENT 39

1. The input impedance of the inverting input in Fig. 39*a*:
   (a) approaches zero; *(b)* approaches infinity; *(c)* is 1 kΩ; *(d)* equals the z_{in} of the op amp. ( )
2. The output voltage recorded in step 1 is:
   (a) 1 mV; *(b)* 10 mV; *(c)* 100 mV; *(d)* 1 V. ( )
3. When you connect the bottom of the 100-kΩ resistor to the inverting input in Fig. 39*a*, you set up an input current of:
   (a) 0.9 μA; *(b)* 9 μA; *(c)* 90 μA; *(d)* 900 μA. ( )
4. The output voltage recorded in step 4 is closest to:
   (a) 0.9 mV; *(b)* 9 mV; *(c)* 90 mV; *(d)* 900 mV. ( )
5. The input impedance looking into pin 3 in Fig. 39*b*:
   (a) approaches zero; *(b)* approaches infinity; *(c)* is 10 Ω; *(d)* equals the z_{in} of the op amp. ( )
6. The i_{out} recorded in step 5 is approximately:
   (a) 1 μA; *(b)* 10 μA; *(c)* 100 μA; *(d)* 1000 μA. ( )
7. A series input connection:
   (a) always decreases input impedance; *(b)* always increases input impedance; *(c)* sometimes decreases input impedance; *(d)* has no effect. ( )
8. A parallel output connection:
   (a) always decreases output impedance; *(b)* always increases output impedance; *(c)* sometimes decreases input impedance; *(d)* has no effect. ( )

9. Optional: Instructor's question.

10. Optional: Instructor's question.

Name: ______________________ Date: __________

QUESTIONS FOR EXPERIMENT 40

1. The value of i_1 in Table 1 is:
   (a) 1 μA; *(b)* 10 μA; *(c)* 100 μA; *(d)* 1 mA. ( )
2. The current gain in Table 1 is approximately:
   (a) 1; *(b)* 10; *(c)* 100; *(d)* 1000. ( )
3. The value of $i_{out(2)}$ in Table 1 is closest to:
   (a) 20 μA; *(b)* 2 mA; *(c)* 20 mA; *(d)* 200 mA. ( )
4. When i_{in} equals 10 μA in Fig. 40*a*, the voltage across the 100-.Ω resistor is approximately:
   (a) 10 μV; *(b)* 1 mV; *(c)* 10 mV; *(d)* 100 mV. ( )
5. When i_{in} is 1 μA in Fig. 40*a*, the voltage across the 10-kΩ resistor is:
   (a) 10 μV; *(b)* 1 mV; *(c)* 10 mV; *(d)* 100 mV. ( )
6. The value of A_1 recorded in step 8 is approximately:
   (a) 1; *(b)* 10; *(c)* 100; *(d)* 1000. ( )
7. The value of A_2 recorded in step 9 is approximately:
   (a) l; *(b)* 10; *(c)* 100; *(d)* 1000. ( )
8. In step 12 you recorded the peak-to-peak output voltage. The rms value of this output voltage is closest to:
   (a) 0.7 V; *(b)* 0.9 V; *(c)* 1 V; *(d)* 2 V. ( )

9. Optional: Instructor's question.

10. Optional: Instructor's question.

Name: ______________________ Date: __________

QUESTIONS FOR EXPERIMENT 41

1. The value of A_{SP} for an R of 22 kΩ in Table 1 is:
   (a) 11; *(b)* 23; *(c)* 48; *(d)* 101. ( )
2. The risetime for an R of 47 kΩ in Table 1 is closest to:
   (a) 4 μs; *(b)* 8 μs; *(c)* 17 μs; *(d)* 35 μs. ( )
3. The calculated values of T_R (Table 1) may differ from the measured values (Table 2) for various reasons. Which of the following cannot be one of the reasons?
   (a) f_{unity} of 741C may differ from 1 MHz; *(b)* tolerance of resistor R; *(c)* tolerance of the 1-kΩ resistor; *(d)* tolerance of the 2.2-kΩ resistor. ( )
4. In Table 1, which of the following describes the relation between A_{SP} and T_R?
   (a) They are directly proportional; *(b)* they are inversely proportional; *(c)* their product is a constant; *(d)* they are exponentially proportional. ( )
5. The values of Table 1 confirm which of these ideas?
   (a) You get more gain when risetime increases; *(b)* bandwidth shrinks as risetime decreases; *(c)* you get more gain when R decreases; *(d)* you get more bandwidth when gain increases. ( )
6. If R equals zero in Fig. 41, the risetime for an f_{unity} of 1 MHz is:
   (a) 0.35 μs; *(b)* 3.5 μs; *(c)* 35 μs; *(d)* 350 μs. ( )
7. In an SP negative-feedback amplifier, the gain-bandwidth product is constant up to f_{unity} under certain conditions. Which of the following is not one of those conditions?
   (a) Internal amplifier must have one dominant lag network up to f_{unity}; *(b)* amplifier must be operating in the linear region; *(c)* resistances must be linear; *(d)* internal amplifier must saturate on positive peaks and cut off on negative peaks. ( )
8. The typical 741C has an internal bel voltage gain of 100 dB up to a break frequency of 10 Hz. At an input frequency of 2 kHz, the internal bel voltage gain equals:
   (a) 66 dB; *(b)* 60 dB; *(c)* 54 dB; *(d)* 34 dB. ( )

9. Optional: Instructor's question.

10. Optional: Instructor's question.

Name: ______________________________ Date: ____________

QUESTIONS FOR EXPERIMENT 42

1. The frequency in Table 1 for an R of 10 kΩ is:
   (a) 159 Hz; *(b)* 318 Hz; *(c)* 1590 Hz; *(d)* 3180 Hz. ( )
2. The frequency in Table 1 for an R of 2.2 kΩ is:
   (a) 1.59 kHz; *(b)* 3.18 kHz; *(c)* 7.23 kHz; *(d)* 10 kHz. ( )
3. The period in Table 2 for an R of 10 kΩ is closest to:
   (a) 628 μs; *(b)* 296 μs; *(c)* 1.73 ms; *(d)* 6.28 ms. ( )
4. The period T in Table 2 for an R value of 4.7 kΩ is closest to:
   (a) 140 μs; *(b)* 300 μs; *(c)* 620 μs; *(d)* 1000 μs. ( )
5. In step 4, the potentiometer is set to approximately:
   (a) 750 Ω; *(b)* 1.5 kΩ; *(c)* 2 kΩ; *(d)* 3 kΩ. ( )
6. In Fig. 42, a tungsten lamp would normally be used in place of the:
   (a) 1-kΩ resistor; *(b)* potentiometer; *(c)* R resistor; *(d)* 0.01-μF capacitor. ( )
7. The signal arriving at the inverting input is what kind of feedback signal?
   (a) PS; *(b)* positive; *(c)* negative; *(d)* PP. ( )
8. The signal returning to the noninverting input represents what kind of feedback?
   (a) PS; *(b)* positive; *(c)* negative; *(d)* SS. ( )

9. Optional: Instructor's question.

10. Optional: Instructor's question.

Name: ______________________________ Date: ______________

QUESTIONS FOR EXPERIMENT 43

1. Neglecting transistor and stray-wiring capacitance, the equivalent capacitance of the tank circuit in Fig. 43 is:
   (a) 909 pF; *(b)* 1000 pF; *(c)* 0.00909 μF; *(d)* 0.01 μF. ( )
2. The resonant frequency recorded in step 2 is closest to:
   (a) 500 kHz; *(b)* 525 kHz; *(c)* 550 kHz; *(d)* 5 MHz. ( )
3. The period recorded in step 4 is closest to:
   (a) 2 ns; *(b)* 20 ns; *(c)* 0.2 μs; *(d)* 2 μs. ( )
4. Before oscillations start, the collector current is closest to:
   (a) 8 μA; *(b)* 20 μA; *(c)* 1 mA; *(d)* 8 mA. ( )
5. For oscillations to start, the small-signal voltage gain must be greater than approximately:
   (a) 5; *(b)* 10; *(c)* 20; *(d)* 50. ( )
6. In Fig. 43, suppose R_{tank} is 3 kΩ. The small-signal voltage gain would be roughly:
   (a) 20; *(b)* 45; *(c)* 100; *(d)* 150. ( )
7. To double the frequency of oscillation in Fig. 43, you can change the inductor to:
   (a) 25 μH; *(b)* 50 μH; *(c)* 200 μH; *(d)* 400 μH. ( )
8. The approximate value of output dc voltage in Fig. 43 is:
   (a) 5 V; *(b)* 7.5 V; *(c)* 12 V; *(d)* 15 V. ( )

9. Optional: Instructor's question.

10. Optional: Instructor's question.

Name: ______________________ Date: ____________

QUESTIONS FOR EXPERIMENT 44

1. Motorboating is caused by:
   (a) supply impedance; *(b)* interstage coupling; *(c)* long supply lead; *(d)* ground loops. ( )
2. The 10-Ω resistor simulates:
   (a) power-supply impedance; *(b)* interstage coupling; *(c)* ground loops; *(d)* long supply lead. ( )
3. The 4 feet of hookup wire produced:
   (a) ground loops; *(b)* excessive lead inductance;*(c)* interstage coupling; *(d)* low supply voltage. ( )
4. The 1-μF capacitor eliminates high-frequency oscillations caused by:
   (a) excessive supply lead length; *(b)* ground loops;*(c)* interstage coupling; *(d)* supply resistance. ( )
5. Ground loops produce:
   (a) low-frequency oscillations; *(b)* high-frequency oscillations; *(c)* motorboating; *(d)* supply resistance. ( )
6. If the first two stages each have a voltage gain of 120 and the last stage has a voltage gain of 200, the overall gain is:
   (a) 14,400; *(b)* 256,000; *(c)* 2,880,000; *(d)* 5,000,000. ( )
7. Because the voltage gain of the three-stage amplifier is so high, it is easy to:
   (a) avoid oscillations; *(b)* get oscillations: *(c)* build nonoscillating high-gain amplifiers; *(d)* prevent oscillations by not using an input signal. ( )
8. Which of the following cannot stop high-frequency oscillations?
   (a) Shielding; *(b)* increasing the distance between stages; *(c)* using a single-point ground system; *(d)* using a regulated power supply. ( )

9. Optional: Instructor's question.

10. Optional: Instructor's question.

Name: ______________________ Date: __________

QUESTIONS FOR EXPERIMENT 45

1. With $V_3 = 15$ V, increasing V_1 eventually:
   (a) turns the LED off; *(b)* turns the LED on; *(c)* makes the LED dimmer; *(d)* makes the LED brighter. ( )
2. Increasing V_1 does which of the following to the gate current?
   (a) Has no effect; *(b)* decreases it; *(c)* increases it; *(d)* makes it equal to zero. ( )
3. Increasing V_1 when $V_2 = 15$ V causes the LED to:
   (a) stay lit; *(b)* go out; *(c)* come on; *(d)* get dimmer. ( )
4. When V_3 is 15 V and the LED is on, the LED current is closest to:
   (a) 5 mA; *(b)* 10 mA; *(c)* 20 mA; *(d)* 30 mA. ( )
5. Increasing V_3 will do which of the following to the LED?
   (a) Make it dimmer; *(b)* make it brighter; *(c)* has no effect; *(d)* shut it off. ( )
6. Once the SCR latches, it stays latched until:
   (a) V_1 is reduced to zero; *(b)* V_3 is increased to 15 V; *(c)* V_1 is increased to 15 V; *(d)* the SCR current drops below the holding current. ( )
7. Allowing 0.7 V for the gate voltage, the trigger or gate current that just causes the LED to come on is closest to:
   (a) 0.1 mA; *(b)* 0.5 mA; *(c)* 10 mA; *(d)* 50 mA. ( )
8. If 2 V is across the LED and 0.7 V across the SCR, the holding current is closest to:
   (a) 0.1 mA; *(b)* 0.5 mA; *(c)* 10 mA; *(d)* 50 mA. ( )

9. Optional: Instructor's question.

10. Optional: Instructor's question.

Name: ______________________________ Date: ____________

QUESTIONS FOR EXPERIMENT 46

1. The value of $A_{SP(\min)}$ in Table 1 is approximately:
   (a) 1; *(b)* 1.2; *(c)* 1.5; *(d)* 1.8. ( )
2. The value of $A_{SP(\max)}$ in Table 1 is closest to:
   (a) 1; *(b)* 1.2; *(c)* 1.5; *(d)* 1.8. ( )
3. The $V_{L(\min)}$ in Table 1 is nearest:
   (a) 8.5 V; *(b)* 10 V; *(c)* 11.2 V; *(d)* 12.5 V. ( )
4. The $V_{L(\max)}$ in Table 1 is closest to:
   (a) 8.5 V; *(b)* 10 V; *(c)* 11.2 V; *(d)* 12.5 V. ( )
5. The $I_{\max}$ of Table 1 is:
   (a) 10 mA; *(b)* 20 mA; *(c)* 30 mA; *(d)* 40 mA. ( )
6. The $V_{L(\min)}$ and $V_{L(\max)}$ of Table 2 may differ from those of Table 1. These differences are not caused by:
   (a) tolerance of zener voltage; *(b)* V_{BE} of Q_1 different from 0.7 V; *(c)* Q_3 turn-on; *(d)* tolerance of potentiometer. ( )
7. Using the information in Table 2, the percent change in load voltage caused by changes in V_S is closest to:
   (a) 0; *(b)* 1%; *(c)* 10%; *(d)* 20%. ( )
8. With the voltages measured in step 11, the current through the 2.2-kΩ resistor is closest to:
   (a) 50 μA; *(b)* 150 μA; *(c)* 1 mA; *(d)* 4 mA. ( )

9. Optional: Instructor's question.

10. Optional: Instructor's question.

Name: ______________________________ **Date:** ____________

QUESTIONS FOR EXPERIMENT 47

1. The minimum input voltage for which the regulator works is closest to:
   (a) 1 V; *(b)* 5 V; *(c)* 10 V; *(d)* 15 V. ( )
2. In Table 1, input voltages greater than 10 V produce output voltages that are:
   (a) equal to input voltages; *(b)* poorly regulated; *(c)* very well regulated; *(d)* distorted. ( )
3. With respect to the input ripple, the regulator has:
   (a) no effect; *(b)* amplification; *(c)* attenuation; *(d)* distortion. ( )
4. A good IC regulator should have:
   (a) high ripple rejection; *(b)* poor regulation; *(c)* low output current; *(d)* distortion. ( )
5. In Fig. 47*c*, output voltage V_{OUT} is related to R_2 as follows:
   (a) no relation; *(b)* increases when R_2 increases; *(c)* decreases when R_2 increases; *(d)* constant. ( )
6. In Fig. 47*c*, I_{OUT} is related to R_2 as follows:
   (a) proportional; *(b)* increases when R_2 increases; *(c)* decreases when R_2 increases; *(d)* approximately constant. ( )
7. The output current in Fig. 47*c* is closest to:
   (a) 10 mA; *(b)* 25 mA; *(c)* 55 mA; *(d)* 85 mA. ( )
8. When R_2 is 100 Ω in Fig. 47*c*, the output voltage is closest to:
   (a) 2 V; *(b)* 6 V; *(c)* 8 V; *(d)* 14 V. ( )

9. Optional: Instructor's question.

10. Optional: Instructor's question.

Name: ______________________________ Date: ____________

QUESTIONS FOR EXPERIMENT 48

1. The green LED comes on when the input is positive in Fig. 48*a* because:
   (a) the noninverting input is used; *(b)* the inverting input is used; *(c)* the potentiometer delivers a negative input; *(d)* the output is negative. ( )
2. The red LED lights when:
   (a) it's reverse-biased; *(b)* the output is negative; *(c)* the output is positive; *(d)* the input is positive. ( )
3. The voltage gain of Fig. 48*b* is:
   (a) 1; *(b)* 10; *(c)* 50; *(d)* 100. ( )
4. The output of Fig. 48*b*:
   (a) follows the input; *(b)* is in phase with the input; *(c)* is out of phase with the input; *(d)* is smaller than the input. ( )
5. The input impedance of Fig. 48*b* is:
   (a) 990 Ω; *(b)* 1 k Ω; *(c)* 10 k Ω; *(d)* 11 k Ω. ( )
6. To change the voltage gain of Fig. 48*b* to 20, we can change:
   (a) R_1 to 50 Ω; *(b)* R_1 to 5 k Ω; *(c)* R_2 to 2 k Ω; *(d)* R_2 to 20 k Ω. ( )
7. In Fig. 48*c*, the current through *R* is approximately:
   (a) 1 mA; *(b)* 5 mA; *(c)* 10 mA; *(d)* 15 mA. ( )
8. To change the current through *R* in Fig. 48*c*, we can change:
   (a) the negative supply voltage; *(b)* the value of *R*; *(c)* the input voltage to pin 7; *(d)* the 1-kΩ resistor. ( )

9. Optional: Instructor's question.

10. Optional: Instructor's question.

Name: ______________________________ Date: ____________

QUESTIONS FOR EXPERIMENT 49

1. The diode offset voltage has the following effect in Fig. 49*a*: *(a)* almost none; *(b)* reduces output by 0.7 V; *(c)* reduces output by 1 V; *(d)* half-wave rectifies the signal. ( )
2. The peak input signal in Fig. 49*a* with a 1-V peak output is: *(a)* 1 V; *(b)* 1.7 V; *(c)* 2.4 V; *(d)* 5 V. ( )
3. When an input signal with a peak of 100 mV drives the peak detector of Fig. 49*b*, the dc output is closest to: *(a)* 100 mV; *(b)* 150 mV; *(c)* 200 mV; *(d)* 300 mV. ( )
4. The circuit of Fig. 49*c* is a: *(a)* peak rectifier; *(b)* half-wave rectifier; *(c)* clamper; *(d)* clipper. ( )
5. The circuit of Fig. 49*d* is a: *(a)* positive clipper; *(b)* negative clipper; *(c)* positive clamper; *(d)* negative clamper. ( )
6. One advantage of active diode circuits is: *(a)* big voltage gain; *(b)* better high-frequency response; *(c)* the need for power supplies; *(d)* the effects of diode offset are greatly reduced. ( )
7. The discharging time constant in Fig. 49*b* is: *(a)* 470 μs; *(b)* 47 ms; *(c)* 0.047 s; *(d)* 470 ms. ( )
8. If the diode of Fig. 49*d* were reversed, the circuit would become a: *(a)* negative clipper; *(b)* positive clipper; *(c)* negative clamper; *(d)* positive clamper. ( )

9. Optional: Instructor's question.

10. Optional: Instructor's question.

Name: ____________________ Date: __________

QUESTIONS FOR EXPERIMENT 50

1. The circuit of Fig. 50*a* is a:
   (a) low-pass filter; *(b)* high-pass filter; *(c)* active diode circuit; *(d)* Miller integrator. ( )
2. The break frequency of Fig. 50*a* is closest to:
   (a) 100 Hz; *(b)* 1 kHz; *(c)* 10 kHz; *(d)* 100 kHz. ( )
3. When the input frequency is 10 times the break frequency in Fig. 50*a*, the output signal is down:
   (a) 0 dB; *(b)* 20 dB; *(c)* 40 dB; *(d)* 60 dB. ( )
4. The rolloff rate of the filter shown in Fig. 50*a* is:
   (a) 6 dB per octave; *(b)* 12 dB per octave; *(c)* 20 dB per decade; *(d)* 60 dB per decade. ( )
5. The circuit of Fig. 50*b* is a:
   (a) low-pass filter; *(b)* high-pass filter; *(c)* bandpass filter; *(d)* bandstop filter. ( )
6. The corner frequency in Fig. 50*b* is closest to:
   (a) 100 Hz; *(b)* 1 kHz; *(c)* 10 kHz; *(d)* 100 kHz. ( )
7. When the input frequency is one-tenth of the break frequency in Fig. 50*b*, the output signal is down:
   (a) 20 dB; *(b)* 40 dB; *(c)* 60 dB; *(d)* 80 dB. ( )
8. The rolloff rate of the circuit in Fig. 50*b* is:
   (a) 6 dB/octave; *(b)* 12 dB/decade; *(c)* 40 dB/decade; *(d)* 20 dB/octave. ( )

9. Optional: Instructor's question.

10. Optional: Instructor's question.

Name: ______________________________ **Date:** ____________

QUESTIONS FOR EXPERIMENT 51

1. The notch frequency in step 1 is:
   (a) 795 Hz; *(b)* 1590 Hz; *(c)* 3180 Hz; *(d)* 15.9 kHz. ( )
2. The true notch frequency in step 5 may differ from the calculated notch frequency of step 1 because:
   (a) the power supply is not exactly 25 V; *(b)* the oscilloscope loads the filter; *(c)* of tolerance of parts. ( )
3. If the JFET amplifier of Fig. 51 produces only square-law distortion, the output of the filter after notching out the fundamental is ideally:
   (a) fundamental only; *(b)* second harmonic only; *(c)* second plus higher harmonics; *(d)* fundamental plus higher harmonics. ( )
4. Ideally, the spectrum of the signal into the notch filter of Fig. 51 contains a fundamental plus:
   (a) second harmonic; *(b)* second and third harmonics; *(c)* all harmonics; *(d)* dc component plus second harmonic. ( )
5. In step 11, the distorted wave has half-wave symmetry. This is:
   (a) true; *(b)* false. ( )
6. The harmonic distortion values of Table 1:
   (a) increase with signal level; *(b)* decrease with signal level; *(c)* sometimes increase, sometimes decrease with level. ( )
7. Table 1 proves the amplifier becomes more linear at:
   (a) lower signal levels; *(b)* higher signal levels; *(c)* intermediate signal levels. ( )
8. If you drive the JFET amplifier of Fig. 51 into cutoff, you will get all harmonics because the operation will no longer be square law. This being the case, the third harmonic coming out of the notch filter has a frequency of approximately:
   (a) 1600 Hz; *(b)* 3200 Hz; *(c)* 4800 Hz; *(d)* 6400 Hz. ( )

9. Optional: Instructor's question.

10. Optional: Instructor's question.

Name: ______________________________ Date: __________

QUESTIONS FOR EXPERIMENT 52

1. The lowest frequency in step 1 is:
   (a) 1 kHz; *(b)* 2 kHz; *(c)* 49 kHz; *(d)* 50 kHz. ( )
2. The highest frequency in step 1 is:
   (a) 99 kHz; *(b)* 101 kHz; *(c)* 151 kHz; *(d)* 201 kHz. ( )
3. The lowest frequency recorded in step 3 is:
   (a) 1 kHz; *(b)* 3 kHz; *(c)* 53 kHz; *(d)* 103 kHz. ( )
4. Which of the following frequencies was not recorded in step 4?
   (a) 1 kHz; *(b)* 2 kHz; *(c)* 3 kHz; *(d)* 4 kHz. ( )
5. Based on steps 6 and 11, the conversion voltage gain in decibels is closest to:
   (a) −20 dB; *(b)* 0 dB; *(c)* 20 dB; *(d)* 40 dB. ( )
6. If the final output signal of Fig. 52 looks like a distorted 1-kHz sine wave, it's because the final output:
   (a) should be a square wave; *(b)* is too weak; *(c)* contains harmonics of 1 kHz; *(d)* should be a spectrum. ( )
7. The ripple on the 1-kHz signal seen at point *B* is:
   (a) 2 kHz; *(b)* 3 kHz; *(c)* 10 kHz; *(d)* 50 kHz. ( )
8. If an ideal FET mixer were used in place of the bipolar mixer, which of these would not happen in Fig. 52?
   (a) The final output would contain 2 kHz; *(b)* the spectrum would contain only group-1 frequencies; *(c)* the final output would contain only 1 kHz plus high-frequency ripple. ( )

9. Optional: Instructor's question.

10. Optional: Instructor's question.

Name: ______________________ Date: __________

QUESTIONS FOR EXPERIMENT 53

1. The ideal voltage gain recorded in step 1 is approximately:
   (a) 10; *(b)* 26; *(c)* 150; *(d)* 400. ( )
2. The maximum voltage gain recorded in step 2 is approximately:
   (a) 8; *(b)* 25; *(c)* 45; *(d)* 65. ( )
3. The minimum voltage gain in step 2 is approximately:
   (a) 8; *(b)* 25; *(c)* 45; *(d)* 65. ( )
4. The percent modulation in step 3 is nearest:
   (a) 10%; *(b)* 30%; *(c)* 70%; *(d)* 100%. ( )
5. The quiescent voltage gain recorded in step 7 may differ from the ideal voltage gain found in step 1. Which of the following is not a possible reason?
   (a) The upper break frequency of the amplifier is less than 500 kHz; *(b)* the r'_e of the transistor may be larger than calculated; *(c)* the 0.001-μF capacitor may look like a short to the modulating signal; *(d)* tolerance of resistors. ( )
6. The percent modulation recorded in step 10 is nearest:
   (a) 1%; *(b)* 3%; *(c)* 10%; *(d)* 30%. ( )
7. The upper side frequency in Fig. 53 is:
   (a) 200 Hz; *(b)* 499,800 Hz; *(c)* 500 kHz; *(d)* 500,200 Hz. ( )
8. The output signal seen in step 8 is an example of:
   (a) DSB-SC; *(b)* DSB-TC; *(c)* SSB-SC; *(d)* SSB-TC. ( )

9. Optional: Instructor's question.

10. Optional: Instructor's question.

Name: ____________________ Date: __________

QUESTIONS FOR EXPERIMENT 54

1. In Fig. 54, the collector sees an ac load resistance of approximately: *(a)* 500 Ω; *(b)* 1 kΩ; *(c)* 2 kΩ; *(d)* 10 kΩ. ( )
2. If β is 100, the dc collector current is approximately: *(a)* 1 mA; *(b)* 6 mA; *(c)* 12 mA; *(d)* 50 mA. ( )
3. The discharging time constant of the envelope detector in Fig. 54 is: *(a)* 1 μs; *(b)* 5 μs; *(c)* 100 μs; *(d)* 500 μs. ( )
4. The $f_{y(\max)}$ recorded in step 1 is closest to: *(a)* 1 kHz; *(b)* 3 kHz; *(c)* 4 kHz; *(d)* 5 kHz. ( )
5. The input spectrum to the envelope detector in Fig. 54 does not contain a(n): *(a)* dc component; *(b)* lower sideband; *(c)* upper sideband; *(d)* carrier. ( )
6. The output spectrum of the envelope detector contains mostly: *(a)* carrier; *(b)* lower sideband; *(c)* upper sideband; *(d)* modulating signal. ( )
7. If the modulating signal has a frequency of 1 kHz and the modulation is 50%, the *RC* time constant of the envelope detector should be no larger than: *(a)* 159 μs; *(b)* 318 μs; *(c)* 1.59 μs; *(d)* 3.18 μs. ( )
8. If the capacitor in the envelope detector is too small, the detected output will have: *(a)* too much carrier ripple; *(b)* not enough carrier ripple; *(c)* not enough modulating signal; *(d)* too much modulating signal. ( )

9. Optional: Instructor's question.

10. Optional: Instructor's question.

appendix

PARTS AND EQUIPMENT

RESISTORS (all ½ W)

Quantity	Description
1	10 Ω
1	33 Ω
1	47 Ω
2	100 Ω
2	150 Ω
1	180 Ω
2	220 Ω
2	330 Ω
4	470 Ω
2	680 Ω
3	1 kΩ
2	2.2 kΩ
1	3.3 kΩ
2	4.7 kΩ
1	6.8 kΩ
4	10 kΩ
2	15 kΩ
1	20 kΩ
2	22 kΩ
2	33 kΩ
1	47 kΩ
1	68 kΩ
2	100 kΩ
3	220 kΩ
1	470 kΩ
1	1 MΩ
1	10 MΩ

POTENTIOMETERS

Quantity	Description
1	5 kΩ
1	50 kΩ
1	100 kΩ
1	1 MΩ

CAPACITORS (25-V rating or better)

Quantity	Description
2	100 pF
2	220 pF
2	1000 pF
1	2000 pF
2	0.01 μF
1	0.02 μF
1	0.022 μF
1	0.1 μF
2	1 μF
1	10 μF
2	47 μF
1	100 μF
2	470 μF

DIODES

Quantity	Description
2	1N753
2	1N914
4	1N4001
1	TIL221
1	TIL222

TRANSISTORS

Quantity	Description
3	2N3904
1	2N3906
1	MPF102
2	2N4444

INTEGRATED CIRCUITS

Quantity	Description
1	LM318C (DIL-8 or TO-5)
1	LM340-8
1	LM741C (DIL-8 or TO-5)

EQUIPMENT

Quantity	Description
1	Ac millivoltmeter
1	Audio generator (two are needed for Exp. 52)
1	DVM (not necessary but desirable)
1	Oscilloscope
2	Power supply (adjustable from at least 1 to 15 V)
1	Power supply (adjustable from at least 15 to 25 V)
1	Sine/square wave generator
1	VOM

MISCELLANEOUS

Quantity	Description
1	Decade resistance box (not necessary but desirable)
1	Inductor, 100 μH
1	Transformer, 12.6 V CT with fused line cord (Triad F-25X or equivalent)

DATA SHEETS

Designers▲Data Sheet

500-MILLIWATT HERMETICALLY SEALED GLASS SILICON ZENER DIODES

- Complete Voltage Range – 2.4 to 91 Volts
- DO-35 Package – Smaller than Conventional DO-7 Package
- Double Slug Type Construction
- Metallurgically Bonded Construction
- Nitride Passivated Die

Designer's Data for "Worst Case" Conditions

The Designers▲ Data sheets permit the design of most circuits entirely from the information presented. Limit curves – representing boundaries on device characteristics – are given to facilitate "worst case" design.

1N746 thru 1N759
1N957 thru 1N984
1N4370 thru 1N4372

GLASS ZENER DIODES

500 MILLIWATTS

2.4-91 VOLTS

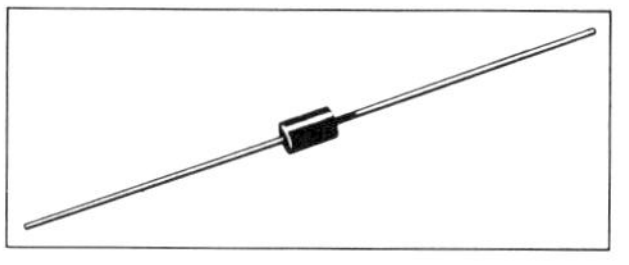

MAXIMUM RATINGS

Rating	Symbol	Value	Unit
DC Power Dissipation @ $T_L \leq 50^oC$, Lead Length = 3/8"	P_D		
*JEDEC Registration		400	mW
*Derate above $T_L = 50^oC$		3.2	mW/^oC
Motorola Device Ratings		500	mW
Derate above $T_L = 50^oC$		3.33	mW/^oC
Operating and Storage Junction Temperature Range	T_J, T_{stg}		^oC
*JEDEC Registration		-65 to +175	
Motorola Device Ratings		-65 to +200	

*Indicates JEDEC Registered Data.

MECHANICAL CHARACTERISTICS

CASE: Double slug type, hermetically sealed glass

MAXIMUM LEAD TEMPERATURE FOR SOLDERING PURPOSES: 230°C, 1/16" from case for 10 seconds

FINISH: All external surfaces are corrosion resistant with readily solderable leads.

POLARITY: Cathode indicated by color band. When operated in zener mode, cathode will be positive with respect to anode.

MOUNTING POSITION: Any

STEADY STATE POWER DERATING

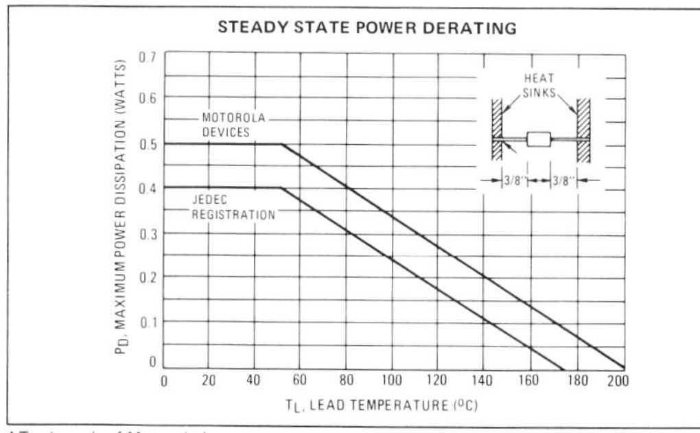

▲Trademark of Motorola Inc.

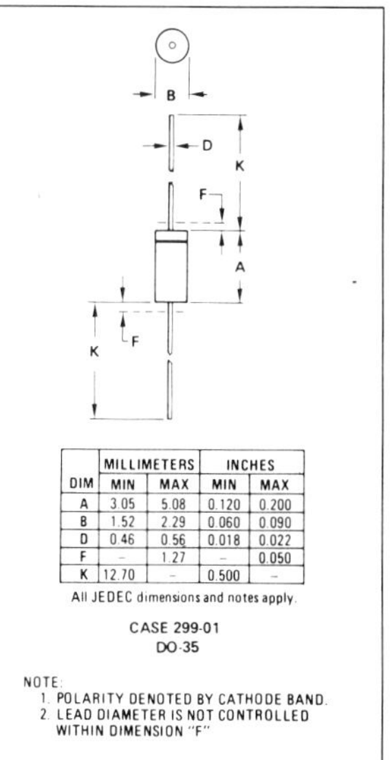

DIM	MILLIMETERS MIN	MILLIMETERS MAX	INCHES MIN	INCHES MAX
A	3.05	5.08	0.120	0.200
B	1.52	2.29	0.060	0.090
D	0.46	0.56	0.018	0.022
F	–	1.27	–	0.050
K	12.70	–	0.500	–

All JEDEC dimensions and notes apply.

CASE 299-01
DO-35

NOTE:
1. POLARITY DENOTED BY CATHODE BAND.
2. LEAD DIAMETER IS NOT CONTROLLED WITHIN DIMENSION "F"

DS 7021 R2

ELECTRICAL CHARACTERISTICS ($T_A = 25^oC$, $V_F = 1.5$ V max at 200 mA for all types)

Type Number (Note 1)	Nominal Zener Voltage V_Z @ I_{ZT} (Note 2) Volts	Test Current I_{ZT} mA	Maximum Zener Impedance Z_{ZT} @ I_{ZT} (Note 3) Ohms	*Maximum DC Zener Current I_{ZM} (Note 4) mA		Maximum Reverse Leakage Current $T_A = 25^oC$ I_R @ $V_R = 1$ V μA	$T_A = 150^oC$ I_R @ $V_R = 1$ V μA
1N4370	2.4	20	30	150	190	100	200
1N4371	2.7	20	30	135	165	75	150
1N4372	3.0	20	29	120	150	50	100
1N746	3.3	20	28	110	135	10	30
1N747	3.6	20	24	100	125	10	30
1N748	3.9	20	23	95	115	10	30
1N749	4.3	20	22	85	105	2	30
1N750	4.7	20	19	75	95	2	30
1N751	5.1	20	17	70	85	1	20
1N752	5.6	20	11	65	80	1	20
1N753	6.2	20	7	60	70	0.1	20
1N754	6.8	20	5	55	65	0.1	20
1N755	7.5	20	6	50	60	0.1	20
1N756	8.2	20	8	45	55	0.1	20
1N757	9.1	20	10	40	50	0.1	20
1N758	10	20	17	35	45	0.1	20
1N759	12	20	30	30	35	0.1	20

Type Number (Note 1)	Nominal Zener Voltage V_Z (Note 2) Volts	Test Current I_{ZT} mA	Maximum Zener Impedance (Note 3) Z_{ZT} @ I_{ZT} Ohms	Z_{ZK} @ I_{ZK} Ohms	I_{ZK} mA	*Maximum DC Zener Current I_{ZM} (Note 4) mA		Maximum Reverse Current I_R Maximum μA	Test Voltage Vdc V_R 5%	10%
1N957	6.8	18.5	4.5	700	1.0	47	61	150	5.2	4.9
1N958	7.5	16.5	5.5	700	0.5	42	55	75	5.7	5.4
1N959	8.2	15	6.5	700	0.5	38	50	50	6.2	5.9
1N960	9.1	14	7.5	700	0.5	35	45	25	6.9	6.6
1N961	10	12.5	8.5	700	0.25	32	41	10	7.6	7.2
1N962	11	11.5	9.5	700	0.25	28	37	5	8.4	8.0
1N963	12	10.5	11.5	700	0.25	26	34	5	9.1	8.6
1N964	13	9.5	13	700	0.25	24	32	5	9.9	9.4
1N965	15	8.5	16	700	0.25	21	27	5	11.4	10.8
1N966	16	7.8	17	700	0.25	19	37	5	12.2	11.5
1N967	18	7.0	21	750	0.25	17	23	5	13.7	13.0
1N968	20	6.2	25	750	0.25	15	20	5	15.2	14.4
1N969	22	5.6	29	750	0.25	14	18	5	16.7	15.8
1N970	24	5.2	33	750	0.25	13	17	5	18.2	17.3
1N971	27	4.6	41	750	0.25	11	15	5	20.6	19.4
1N972	30	4.2	49	1000	0.25	10	13	5	22.8	21.6
1N973	33	3.8	58	1000	0.25	9.2	12	5	25.1	23.8
1N974	36	3.4	70	1000	0.25	8.5	11	5	27.4	25.9
1N975	39	3.2	80	1000	0.25	7.8	10	5	29.7	28.1
1N976	43	3.0	93	1500	0.25	7.0	9.6	5	32.7	31.0
1N977	47	2.7	105	1500	0.25	6.4	8.8	5	35.8	33.8
1N978	51	2.5	125	1500	0.25	5.9	8.1	5	38.8	36.7
1N979	56	2.2	150	2000	0.25	5.4	7.4	5	42.6	40.3
1N980	62	2.0	185	2000	0.25	4.9	6.7	5	47.1	44.6
1N981	68	1.8	230	2000	0.25	4.5	6.1	5	51.7	49.0
1N982	75	1.7	270	2000	0.25	1.0	5.5	5	56.0	54.0
1N983	82	1.5	330	3000	0.25	3.7	5.0	5	62.2	59.0
1N984	91	1.4	400	3000	0.25	3.3	4.5	5	69.2	65.5

*Left column based upon JEDEC Registration, right column based upon Motorola rating.

NOTE 1. TOLERANCE AND VOLTAGE DESIGNATION

Tolerance Designation

The type numbers shown have tolerance designations as follows:

1N4370 series: ±10%, suffix A for ±5% units.

1N746 series: ±10%, suffix A for ±5% units.

1N957 series: ±20%, suffix A for ±10% units, suffix B for ±5% units.

NOTE 2. ZENER VOLTAGE (V_Z) MEASUREMENT

Nominal zener voltage is measured with the device junction in thermal equilibrium at the lead temperature of $30^oC \pm 1^oC$ and 3/8" lead length.

NOTE 3. ZENER IMPEDANCE (Z_Z) DERIVATION

Z_{ZT} and Z_{ZK} are measured by dividing the ac voltage drop across the device by the ac current applied. The specified limits are for $I_Z(ac) = 0.1\ I_Z(dc)$ with the ac frequency = 60 Hz.

NOTE 4. MAXIMUM ZENER CURRENT RATINGS (I_{ZM})

Maximum zener current ratings are based on the maximum voltage of a 10% 1N746 type unit or a 20% 1N957 type unit. For closer tolerance units (10% or 5%) or units where the actual zener voltage (V_Z) is known at the operating point, the maximum zener current may be increased and is limited by the derating curve.

1N914

ULTRA FAST PLANAR* DIODE

MAXIMUM RATINGS (25°C) [Note 1]

WIV	Working Inverse Voltage	20 V
I_O	Average Rectified Current	50 mA
I_F	Forward Current Steady State d.c.	75 mA
I_F	Recurrent Peak Forward Current	150 mA
$i_{f(surge)}$	Peak Forward Surge Current Pulse Width of 1 Second	500 mA
$i_{f(surge)}$	Peak Foward Surge Current Pulse Width of 1 μs	2000 mA
P	Power Dissipation	250 mW
P	Power Dissipation	100 mW at 125°C
T_A	Operating Temperature	−65° to +175°C
T_{stg}	Storage Temperature, Ambient	−65° to +175°C

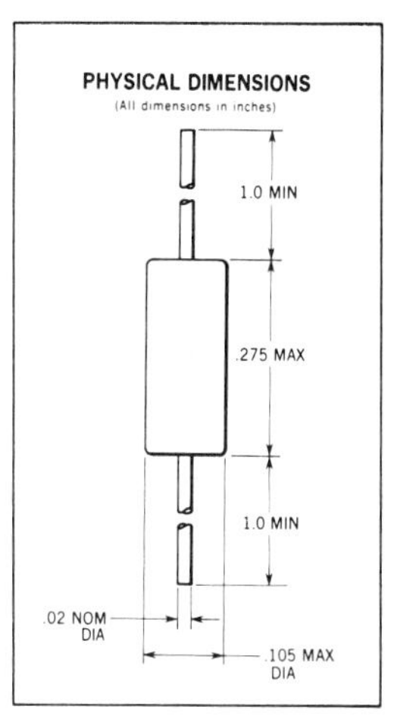

ELECTRICAL SPECIFICATIONS (25°C unless otherwise noted)

SYMBOL	CHARACTERISTIC	MIN.	TYP.	MAX.	TEST CONDITIONS
V_F	Forward Voltage			1.0 V	$I_F = 10$ mA
I_R	Reverse Current			25 NA	$V_R = -20$ V
I_R	Reverse Current (150°C)			50 μA	$V_R = -20$ V
BV	Breakdown Voltage	75 V		50 μA	$I_R = -25$ μA
BV	Breakdown Voltage	100V			$I_R = 100$ μA
t_{rr} (Note 2)	Reverse Recovery Time			4.0 ns	$I_F = 10$ mA $V_R = 6$ V
V_F (Note 5)	Peak forward recovery voltage			2.5 V	$I_F = 50$ mA pulse
C_O (Note 3)	Capacitance			4.0 pF	$V_R = 0$V f = 1
R_E (Note 4)	Rectification Efficiency 45%				f = 100 MHz
ΔV_F/°C	Forward Voltage Temperature Coefficient		−1.8mV/°C		

*Planar is a patented Fairchild process.

NOTES:

(1) The maximum ratings are limiting values above which life or satisfactory performance may be impaired.
(2) Recovery to 1 mA.
(3) Capacitance as measured on Boonton Electronic Corporation Model No. 75A-S8 Capacitance Bridge or equivalent.
(4) Rectification efficiency is defined as the ratio of D.C. load voltage to peak rf input voltage to the detector circuit, measured with 2.0 V r.m.s. input to the circuit. Load resistance 5k ohms, load capacitance 20 pF.
(5) Pulse width = 0.1 μs; Rise time of pulse equal to or less than 25 ns. Repetition rate 5 - 100 kHz.

313 FAIRCHILD DRIVE, MOUNTAIN VIEW, CALIFORNIA, (415) 962-5011, TWX: 910-379-6435

1N4001 thru 1N4007

Designers▲Data Sheet

"SURMETIC"▲ RECTIFIERS

. . . subminiature size, axial lead mounted rectifiers for general-purpose low-power applications.

Designers Data for "Worst Case" Conditions

The Designers▲ Data Sheets permit the design of most circuits entirely from the information presented. Limit curves — representing boundaries on device characteristics — are given to facilitate "worst case" design.

LEAD MOUNTED SILICON RECTIFIERS

50-1000 VOLTS DIFFUSED JUNCTION

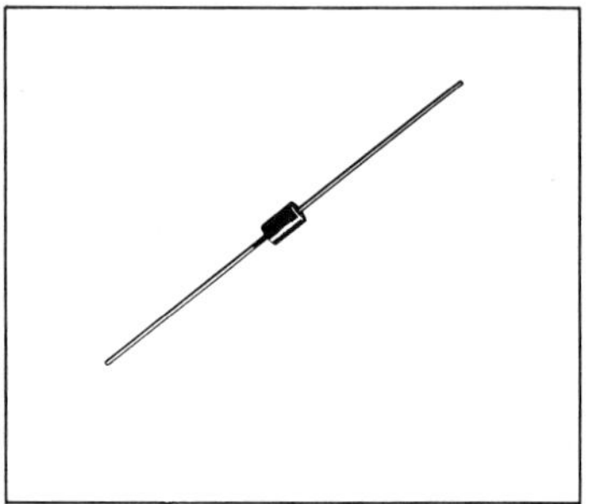

*MAXIMUM RATINGS

Rating	Symbol	1N4001	1N4002	1N4003	1N4004	1N4005	1N4006	1N4007	Unit
Peak Repetitive Reverse Voltage Working Peak Reverse Voltage DC Blocking Voltage	V_{RRM} V_{RWM} V_R	50	100	200	400	600	800	1000	Volts
Non-Repetitive Peak Reverse Voltage (halfwave, single phase, 60 Hz)	V_{RSM}	60	120	240	480	720	1000	1200	Volts
RMS Reverse Voltage	$V_{R(RMS)}$	35	70	140	280	420	560	700	Volts
Average Rectified Forward Current (single phase, resistive load, 60 Hz, see Figure 8, $T_A = 75^oC$)	I_O	1.0							Amp
Non-Repetitive Peak Surge Current (surge applied at rated load conditions, see Figure 2)	I_{FSM}	30 (for 1 cycle)							Amp
Operating and Storage Junction Temperature Range	T_J, T_{stg}	-65 to +175							^oC

*ELECTRICAL CHARACTERISTICS

Characteristic and Conditions	Symbol	Typ	Max	Unit
Maximum Instantaneous Forward Voltage Drop ($i_F = 1.0$ Amp, $T_J = 25^oC$) Figure 1	v_F	0.93	1.1	Volts
Maximum Full-Cycle Average Forward Voltage Drop ($I_O = 1.0$ Amp, $T_L = 75^oC$, 1 inch leads)	$V_{F(AV)}$	–	0.8	Volts
Maximum Reverse Current (rated dc voltage) $T_J = 25^oC$ $T_J = 100^oC$	I_R	0.05 1.0	10 50	μA
Maximum Full-Cycle Average Reverse Current ($I_O = 1.0$ Amp, $T_L = 75^oC$, 1 inch leads)	$I_{R(AV)}$	–	30	μA

*Indicates JEDEC Registered Data.

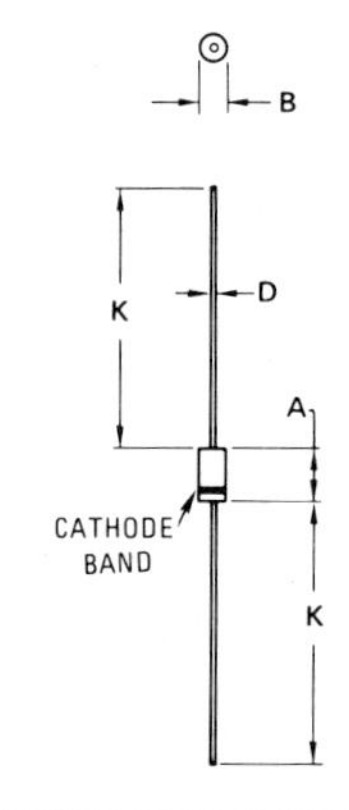

DIM	MILLIMETERS MIN	MILLIMETERS MAX	INCHES MIN	INCHES MAX
A	5.97	6.60	0.235	0.260
B	2.79	3.05	0.110	0.120
D	0.76	0.86	0.030	0.034
K	27.94	–	1.100	–

CASE 59-04

Does Not Conform to DO-41 Outline.

MECHANICAL CHARACTERISTICS

CASE: Void free, Transfer Molded

MAXIMUM LEAD TEMPERATURE FOR SOLDERING PURPOSES: 350°C, 3/8" from case for 10 seconds at 5 lbs. tension

FINISH: All external surfaces are corrosion-resistant, leads are readily solderable

POLARITY: Cathode indicated by color band

WEIGHT: 0.40 Grams (approximately)

▲Trademark of Motorola Inc.

DS 6015 R3

2N3903
2N3904

NPN SILICON ANNULAR♦ TRANSISTORS

. . . designed for general purpose switching and amplifier applications and for complementary circuitry with types 2N3905 and 2N3906.

- High Voltage Ratings – BV_{CEO} = 40 Volts (Min)
- Current Gain Specified from 100 μA to 100 mA
- Complete Switching and Amplifier Specifications
- Low Capacitance – C_{ob} = 4.0 pF (Max)

NPN SILICON SWITCHING & AMPLIFIER TRANSISTORS

MAXIMUM RATINGS

Rating	Symbol	Value	Unit
*Collector-Base Voltage	V_{CB}	60	Vdc
*Collector-Emitter Voltage	V_{CEO}	40	Vdc
*Emitter-Base Voltage	V_{EB}	6.0	Vdc
*Collector Current	I_C	200	mAdc
Total Power Dissipation @ T_A = 60°C	P_D	250	mW
**Total Power Dissipation @ T_A = 25°C Derate above 25°C	P_D	350 2.8	mW mW/°C
**Total Power Dissipation @ T_C = 25°C Derate above 25°C	P_D	1.0 8.0	Watts mW/°C
**Junction Operating Temperature	T_J	150	°C
**Storage Temperature Range	T_{stg}	-55 to +150	°C

THERMAL CHARACTERISTICS

Characteristic	Symbol	Max	Unit
Thermal Resistance, Junction to Ambient	$R_{\theta JA}$	357	°C/W
Thermal Resistance, Junction to Case	$R_{\theta JC}$	125	°C/W

*Indicates JEDEC Registered Data
**Motorola guarantees this data in addition to the JEDEC Registered Data.
♦Annular Semiconductors Patented by Motorola Inc.

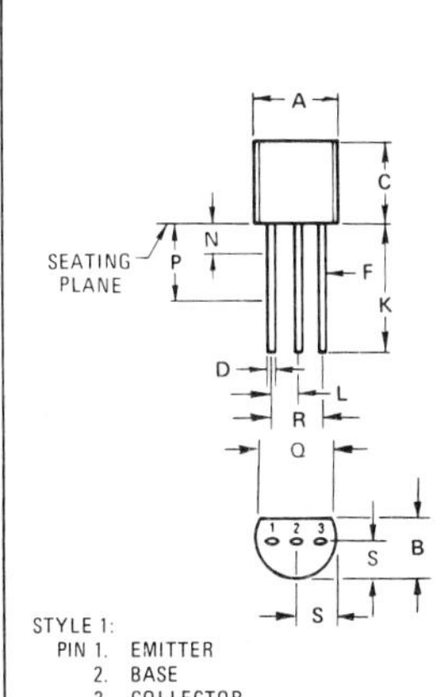

STYLE 1:
PIN 1. EMITTER
2. BASE
3. COLLECTOR

DIM	MILLIMETERS MIN	MILLIMETERS MAX	INCHES MIN	INCHES MAX
A	4.450	5.200	0.175	0.205
B	3.180	4.190	0.125	0.165
C	4.320	5.330	0.170	0.210
D	0.407	0.533	0.016	0.021
F	0.407	0.482	0.016	0.019
K	12.700	–	0.500	–
L	1.150	1.390	0.045	0.055
N	–	1.270	–	0.050
P	6.350	–	0.250	–
Q	3.430	–	0.135	–
R	2.410	2.670	0.095	0.105
S	2.030	2.670	0.080	0.105

CASE 29-02
TO-92

 DS 5127 R2

*ELECTRICAL CHARACTERISTICS (T_A = 25°C unless otherwise noted)

Characteristic		Fig. No.	Symbol	Min	Max	Unit
OFF CHARACTERISTICS						
Collector-Base Breakdown Voltage (I_C = 10 μAdc, I_E = 0)			BV_{CBO}	60	-	Vdc
Collector-Emitter Breakdown Voltage (1) (I_C = 1.0 mAdc, I_B = 0)			BV_{CEO}	40	-	Vdc
Emitter-Base Breakdown Voltage (I_E = 10 μAdc, I_C = 0)			BV_{EBO}	6.0	-	Vdc
Collector Cutoff Current (V_{CE} = 30 Vdc, $V_{EB(off)}$ = 3.0 Vdc)			I_{CEX}	-	50	nAdc
Base Cutoff Current (V_{CE} = 30 Vdc, $V_{EB(off)}$ = 3.0 Vdc)			I_{BL}	-	50	nAdc
ON CHARACTERISTICS						
DC Current Gain (1) (I_C = 0.1 mAdc, V_{CE} = 1.0 Vdc)	2N3903 2N3904	15	h_{FE}	20 40	- -	-
(I_C = 1.0 mAdc, V_{CE} = 1.0 Vdc)	2N3903 2N3904			35 70	- -	
(I_C = 10 mAdc, V_{CE} = 1.0 Vdc)	2N3903 2N3904			50 100	150 300	
(I_C = 50 mAdc, V_{CE} = 1.0 Vdc)	2N3903 2N3904			30 60	- -	
(I_C = 100 mAdc, V_{CE} = 1.0 Vdc)	2N3903 2N3904			15 30	- -	
Collector-Emitter Saturation Voltage (1) (I_C = 10 mAdc, I_B = 1.0 mAdc) (I_C = 50 mAdc, I_B = 5.0 mAdc)		16, 17	$V_{CE(sat)}$	- -	0.2 0.3	Vdc
Base-Emitter Saturation Voltage (1) (I_C = 10 mAdc, I_B = 1.0 mAdc) (I_C = 50 mAdc, I_B = 5.0 mAdc)		17	$V_{BE(sat)}$	0.65 -	0.85 0.95	Vdc
SMALL-SIGNAL CHARACTERISTICS						
Current-Gain—Bandwidth Product (I_C = 10 mAdc, V_{CE} = 20 Vdc, f = 100 MHz)	2N3903 2N3904		f_T	250 300	- -	MHz
Output Capacitance (V_{CB} = 5.0 Vdc, I_E = 0, f = 100 kHz)		3	C_{ob}	-	4.0	pF
Input Capacitance (V_{BE} = 0.5 Vdc, I_C = 0, f = 100 kHz)		3	C_{ib}	-	8.0	pF
Input Impedance (I_C = 1.0 mAdc, V_{CE} = 10 Vdc, f = 1.0 kHz)	2N3903 2N3904	13	h_{ie}	0.5 1.0	8.0 10	k ohms
Voltage Feedback Ratio (I_C = 1.0 mAdc, V_{CE} = 10 Vdc, f = 1.0 kHz)	2N3903 2N3904	14	h_{re}	0.1 0.5	5.0 8.0	X 10^{-4}
Small-Signal Current Gain (I_C = 1.0 mAdc, V_{CE} = 10 Vdc, f = 1.0 kHz)	2N3903 2N3904	11	h_{fe}	50 100	200 400	
Output Admittance (I_C = 1.0 mAdc, V_{CE} = 10 Vdc, f = 1.0 kHz)		12	h_{oe}	1.0	40	μmhos
Noise Figure (I_C = 100 μAdc, V_{CE} = 5.0 Vdc, R_S = 1.0 k ohms, f = 10 Hz to 15.7 kHz)	2N3903 2N3904	9, 10	NF	- -	6.0 5.0	dB
SWITCHING CHARACTERISTICS						
Delay Time	(V_{CC} = 3.0 Vdc, $V_{BE(off)}$ = 0.5 Vdc, I_C = 10 mAdc, I_{B1} = 1.0 mAdc)	1, 5	t_d	-	35	ns
Rise Time		1, 5, 6	t_r	-	35	ns
Storage Time	(V_{CC} = 3.0 Vdc, I_C = 10 mAdc, I_{B1} = I_{B2} = 1.0 mAdc) 2N3903 / 2N3904	2, 7	t_s	- -	175 200	ns
Fall Time		2, 8	t_f	-	50	ns

(1) Pulse Test: Pulse Width = 300μs, Duty Cycle = 2.0%.
*Indicates JEDEC Registered Data

FIGURE 1 – DELAY AND RISE TIME EQUIVALENT TEST CIRCUIT

FIGURE 2 – STORAGE AND FALL TIME EQUIVALENT TEST CIRCUIT

*Total shunt capacitance of test jig and connectors

MOTOROLA Semiconductor Products Inc.

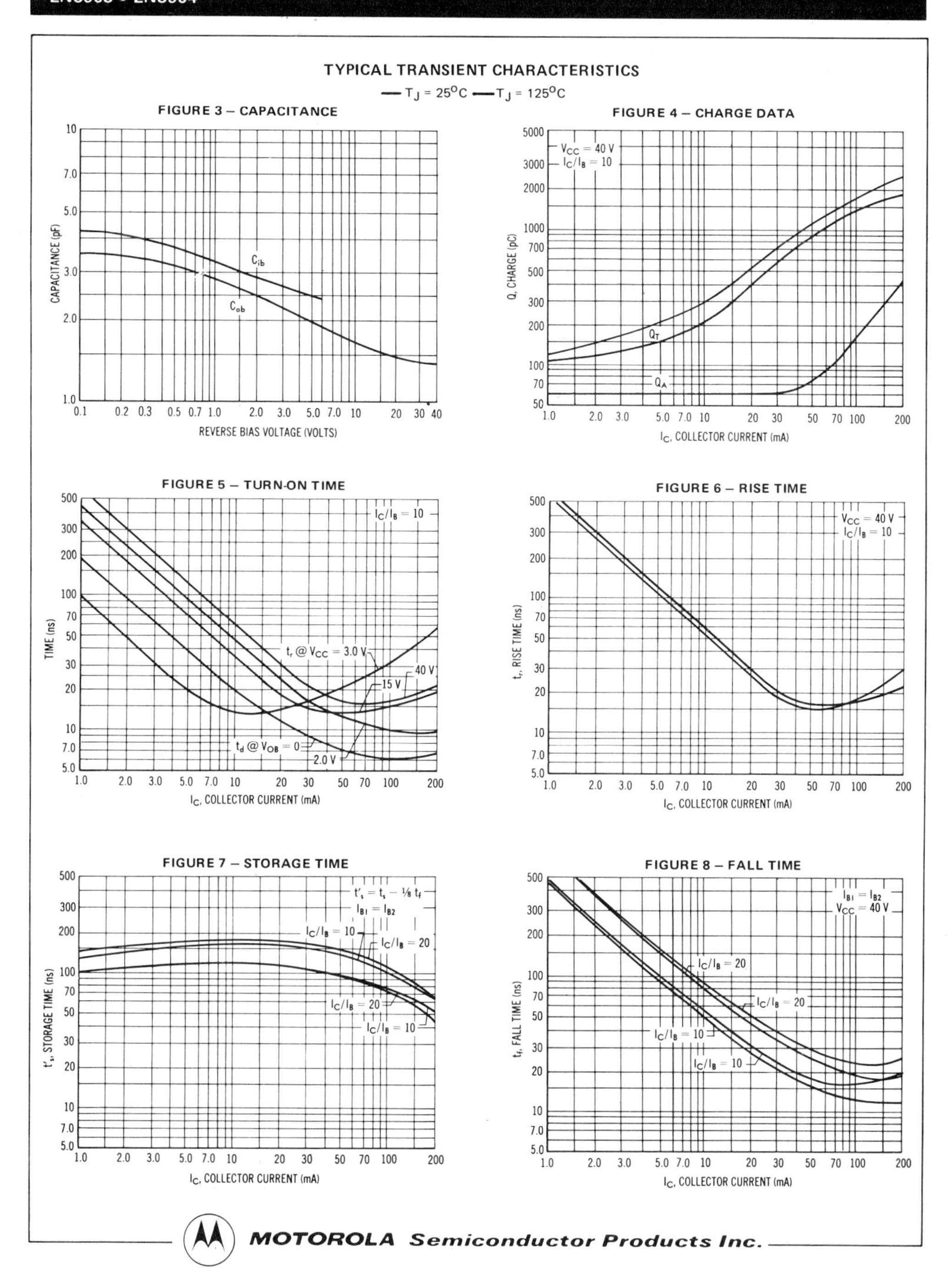
TYPICAL TRANSIENT CHARACTERISTICS
— TJ = 25°C — TJ = 125°C
FIGURE 3 – CAPACITANCE
CAPACITANCE (pF)
Cib
Cob
REVERSE BIAS VOLTAGE (VOLTS)
FIGURE 4 – CHARGE DATA
VCC = 40 V
IC/IB = 10
Q, CHARGE (pC)
QT
QA
IC, COLLECTOR CURRENT (mA)
FIGURE 5 – TURN-ON TIME
IC/IB = 10
TIME (ns)
tr @ VCC = 3.0 V
40 V
15 V
td @ VOB = 0
2.0 V
IC, COLLECTOR CURRENT (mA)
FIGURE 6 – RISE TIME
VCC = 40 V
IC/IB = 10
tr, RISE TIME (ns)
IC, COLLECTOR CURRENT (mA)
FIGURE 7 – STORAGE TIME
t's = ts – 1/8 tf
IB1 = IB2
IC/IB = 10
IC/IB = 20
IC/IB = 20
IC/IB = 10
t's, STORAGE TIME (ns)
IC, COLLECTOR CURRENT (mA)
FIGURE 8 – FALL TIME
IB1 = IB2
VCC = 40 V
IC/IB = 20
IC/IB = 20
IC/IB = 10
IC/IB = 10
tf, FALL TIME (ns)
IC, COLLECTOR CURRENT (mA)
MOTOROLA Semiconductor Products Inc.

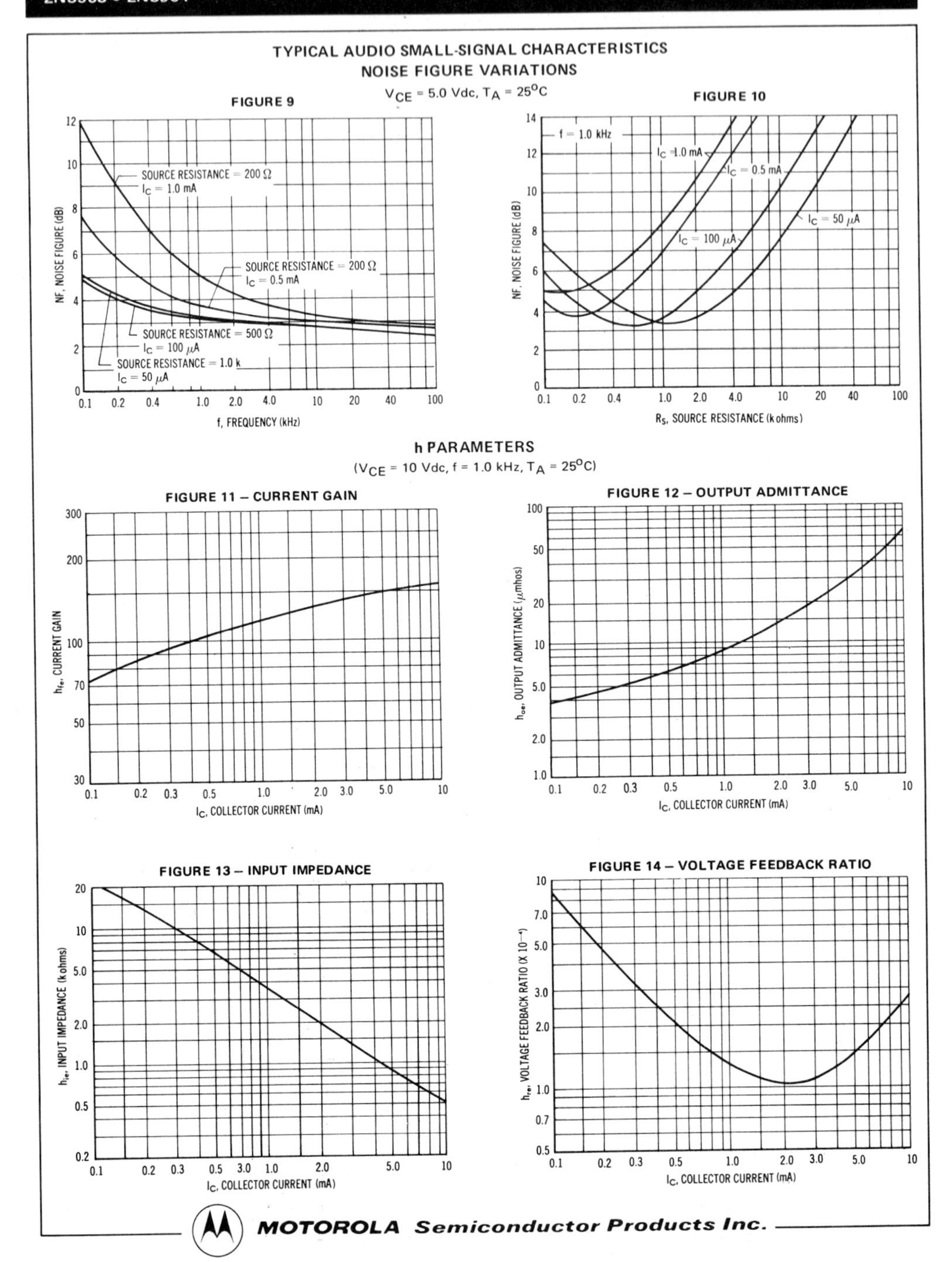

TYPICAL AUDIO SMALL-SIGNAL CHARACTERISTICS
NOISE FIGURE VARIATIONS
V_{CE} = 5.0 Vdc, T_A = 25°C
FIGURE 9
SOURCE RESISTANCE = 200 Ω
I_C = 1.0 mA
SOURCE RESISTANCE = 200 Ω
I_C = 0.5 mA
SOURCE RESISTANCE = 500 Ω
I_C = 100 μA
SOURCE RESISTANCE = 1.0 k
I_C = 50 μA
NF, NOISE FIGURE (dB)
f, FREQUENCY (kHz)
FIGURE 10
f = 1.0 kHz
I_C = 1.0 mA
I_C = 0.5 mA
I_C = 50 μA
I_C = 100 μA
NF, NOISE FIGURE (dB)
R_S, SOURCE RESISTANCE (k ohms)
h PARAMETERS
(V_{CE} = 10 Vdc, f = 1.0 kHz, T_A = 25°C)
FIGURE 11 – CURRENT GAIN
h_{fe}, CURRENT GAIN
I_C, COLLECTOR CURRENT (mA)
FIGURE 12 – OUTPUT ADMITTANCE
h_{oe}, OUTPUT ADMITTANCE (μmhos)
I_C, COLLECTOR CURRENT (mA)
FIGURE 13 – INPUT IMPEDANCE
h_{ie}, INPUT IMPEDANCE (k ohms)
I_C, COLLECTOR CURRENT (mA)
FIGURE 14 – VOLTAGE FEEDBACK RATIO
h_{re}, VOLTAGE FEEDBACK RATIO (X 10^{-4})
I_C, COLLECTOR CURRENT (mA)
MOTOROLA Semiconductor Products Inc.

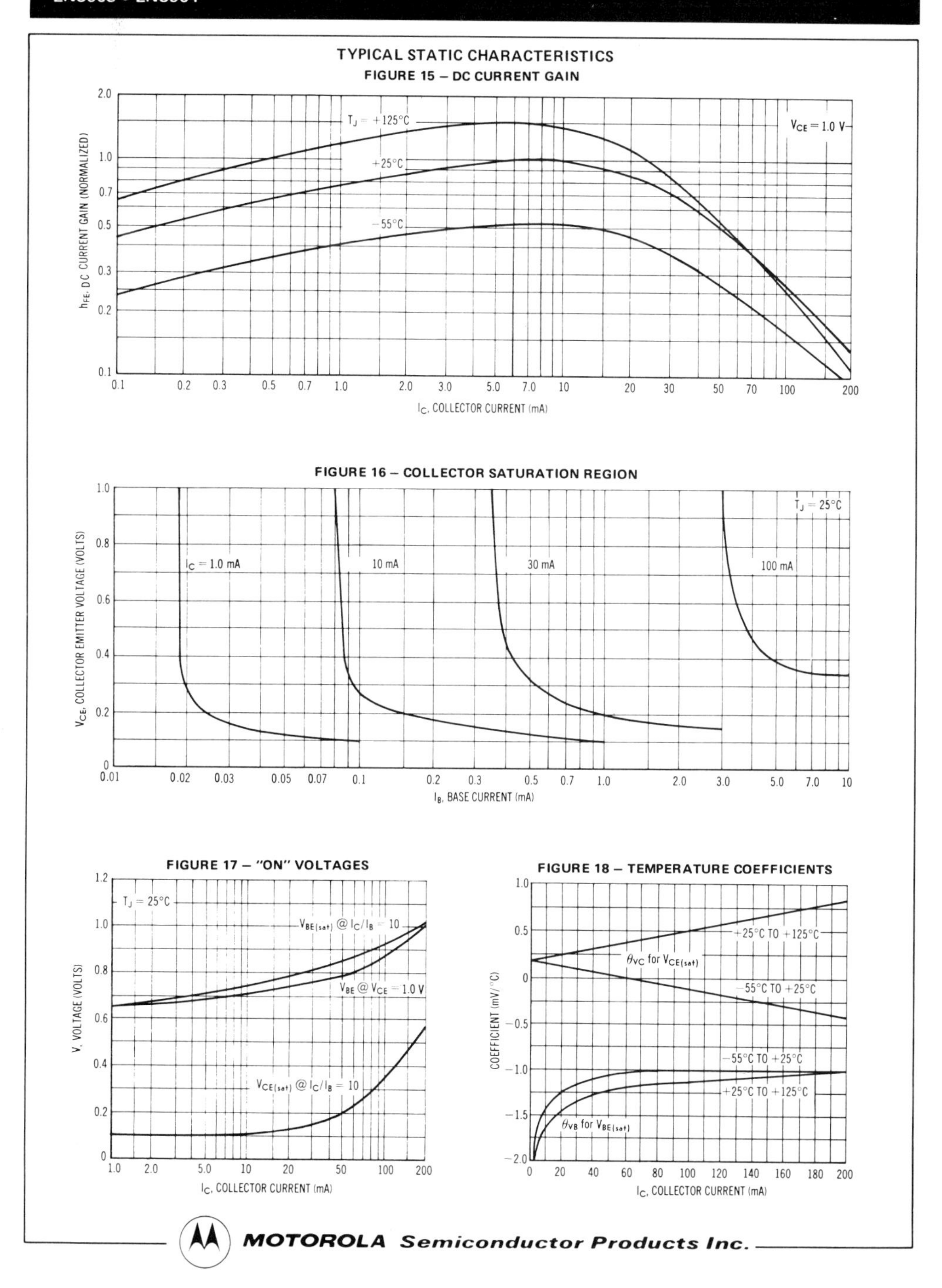

MOTOROLA *Semiconductor Products Inc.*

2N3905
2N3906

PNP SILICON SWITCHING & AMPLIFIER TRANSISTORS

PNP SILICON ANNULAR♦ TRANSISTORS

.... designed for general purpose switching and amplifier applications and for complementary circuitry with types 2N3903 and 2N3904.

- High Voltage Ratings — BV_{CEO} = 40 Volts (Min)
- Current Gain Specified from 100 μA to 100 mA
- Complete Switching and Amplifier Specifications
- Low Capacitance — C_{ob} = 4.5 pF (Max)

SEATING PLANE

STYLE 1:
PIN 1. EMITTER
2. BASE
3. COLLECTOR

	MILLIMETERS		INCHES	
DIM	MIN	MAX	MIN	MAX
A	4.450	5.200	0.175	0.205
B	3.180	4.190	0.125	0.165
C	4.320	5.330	0.170	0.210
D	0.407	0.533	0.016	0.021
F	0.407	0.482	0.016	0.019
K	12.700	–	0.500	–
L	1.150	1.390	0.045	0.055
N	–	1.270	–	0.050
P	6.350	–	0.250	–
Q	3.430	–	0.135	–
R	2.410	2.670	0.095	0.105
S	2.030	2.670	0.080	0.105

CASE 29-02
(TO-92)

*MAXIMUM RATINGS

Rating	Symbol	Value	Unit
Collector-Base Voltage	V_{CB}	40	Vdc
Collector-Emitter Voltage	V_{CEO}	40	Vdc
Emitter-Base Voltage	V_{EB}	5.0	Vdc
Collector Current	I_C	200	mAdc
Total Power Dissipation @ $T_A = 60^oC$	P_D	250	mW
Total Power Dissipation @ $T_A = 25^oC$ Derate above 25^oC	P_D	350 2.8	mW mW/^{o}C
Total Power Dissipation @ $T_C = 25^oC$ Derate above 25^oC	P_D	1.0 8.0	Watt mW/^{o}C
Junction Operating Temperature	T_J	+150	^{o}C
Storage Temperature Range	T_{stg}	-55 to +150	^{o}C

THERMAL CHARACTERISTICS

Characteristic	Symbol	Max	Unit
Thermal Resistance, Junction to Ambient	$R_{\theta JA}$	357	^{o}C/W
Thermal Resistance, Junction to Case	$R_{\theta JC}$	125	^{o}C/W

*Indicates JEDEC Registered Data.
♦Annular semiconductors patented by Motorola Inc.

DS 5128 R2

***ELECTRICAL CHARACTERISTICS** (T_A = 25°C unless otherwise noted.)

Characteristic		Fig. No.	Symbol	Min	Max	Unit
OFF CHARACTERISTICS						
Collector-Base Breakdown Voltage (I_C = 10 μAdc, I_E = 0)			BV_{CBO}	40	–	Vdc
Collector-Emitter Breakdown Voltage (1) (I_C = 1.0 mAdc, I_B = 0)			BV_{CEO}	40	–	Vdc
Emitter-Base Breakdown Voltage (I_E = 10 μAdc, I_C = 0)			BV_{EBO}	5.0	–	Vdc
Collector Cutoff Current (V_{CE} = 30 Vdc, $V_{BE(off)}$ = 3.0 Vdc)			I_{CEX}	–	50	nAdc
Base Cutoff Current (V_{CE} = 30 Vdc, $V_{BE(off)}$ = 3.0 Vdc)			I_{BL}	–	50	nAdc
ON CHARACTERISTICS (1)						
DC Current Gain		15	h_{FE}			
(I_C = 0.1 mAdc, V_{CE} = 1.0 Vdc)	2N3905			30	–	
	2N3906			60	–	
(I_C = 1.0 mAdc, V_{CE} = 1.0 Vdc)	2N3905			40	–	
	2N3906			80	–	
(I_C = 10 mAdc, V_{CE} = 1.0 Vdc)	2N3905			50	150	
	2N3906			100	300	
(I_C = 50 mAdc, V_{CE} = 1.0 Vdc)	2N3905			30	–	
	2N3906			60	–	
(I_C = 100 mAdc, V_{CE} = 1.0 Vdc)	2N3905			15	–	
	2N3906			30	–	
Collector-Emitter Saturation Voltage		16, 17	$V_{CE(sat)}$			Vdc
(I_C = 10 mAdc, I_B = 1.0 mAdc)				–	0.25	
(I_C = 50 mAdc, I_B = 5.0 mAdc)				–	0.4	
Base-Emitter Saturation Voltage		17	$V_{BE(sat)}$			Vdc
(I_C = 10 mAdc, I_B = 1.0 mAdc)				0.65	0.85	
(I_C = 50 mAdc, I_B = 5.0 mAdc)				–	0.95	
SMALL-SIGNAL CHARACTERISTICS						
Current-Gain – Bandwidth Product (I_C = 10 mAdc, V_{CE} = 20 Vdc, f = 100 MHz)	2N3905		f_T	200	–	MHz
	2N3906			250	–	
Output Capacitance (V_{CB} = 5.0 Vdc, I_E = 0, f = 100 kHz)		3	C_{ob}	–	4.5	pF
Input Capacitance (V_{BE} = 0.5 Vdc, I_C = 0, f = 100 kHz)		3	C_{ib}	–	1.0	pF
Input Impedance (I_C = 1.0 mAdc, V_{CE} = 10 Vdc, f = 1.0 kHz)	2N3905	13	h_{ie}	0.5	8.0	k ohms
	2N3906			2.0	12	
Voltage Feedback Ratio (I_C = 1.0 mAdc, V_{CE} = 10 Vdc, f = 1.0 kHz)	2N3905	14	h_{re}	0.1	5.0	X 10^{-4}
	2N3906			1.0	10	
Small-Signal Current Gain (I_C = 1.0 mAdc, V_{CE} = 10 Vdc, f = 1.0 kHz)	2N3905	11	h_{fe}	50	200	–
	2N3906			100	400	
Output Admittance (I_C = 1.0 mAdc, V_{CE} = 10 Vdc, f = 1.0 kHz)	2N3905	12	h_{oe}	1.0	40	μmhos
	2N3906			3.0	60	
Noise Figure (I_C = 100 μAdc, V_{CE} = 5.0 Vdc, R_S = 1.0 k ohm, f = 10 Hz to 15.7 kHz)	2N3905	9, 10	NF	–	5.0	dB
	2N3906			–	4.0	
SWITCHING CHARACTERISTICS						
Delay Time (V_{CC} = 3.0 Vdc, $V_{BE(off)}$ = 0.5 Vdc, I_C = 10 mAdc, I_{B1} = 1.0 mAdc)		1, 5	t_d	–	35	ns
Rise Time (V_{CC} = 3.0 Vdc, $V_{BE(off)}$ = 0.5 Vdc, I_C = 10 mAdc, I_{B1} = 1.0 mAdc)		1, 5, 6	t_r	–	35	ns
Storage Time (V_{CC} = 3.0 Vdc, I_C = 10 mAdc, $I_{B1} = I_{B2}$ = 1.0 mAdc)	2N3905	2, 7	t_s	–	200	ns
	2N3906			–	225	
Fall Time (V_{CC} = 3.0 Vdc, I_C = 10 mAdc, $I_{B1} = I_{B2}$ = 1.0 mAdc)	2N3905	2, 8	t_f	–	60	ns
	2N3906			–	75	

*Indicates JEDEC Registered Data. (1) Pulse Width = 300 μs, Duty Cycle = 2.0 %.

FIGURE 1 – DELAY AND RISE TIME EQUIVALENT TEST CIRCUIT

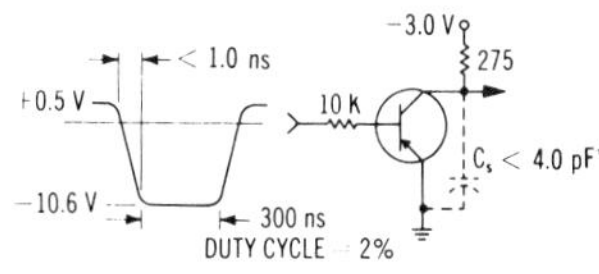

FIGURE 2 – STORAGE AND FALL TIME EQUIVALENT TEST CIRCUIT

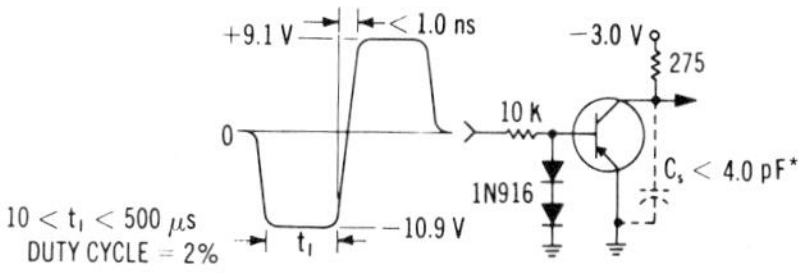

*Total shunt capacitance of test jig and connectors

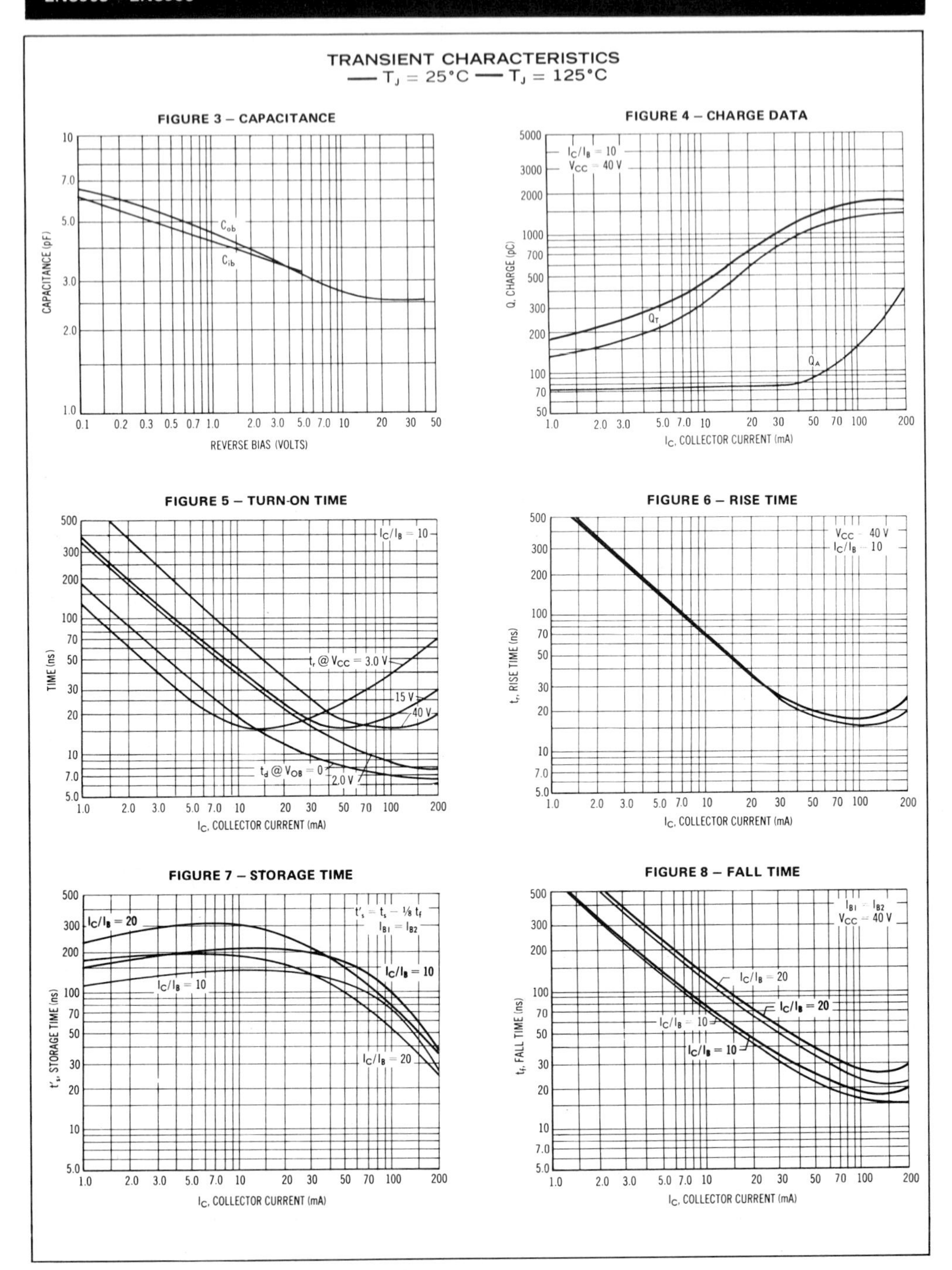
TRANSIENT CHARACTERISTICS
— TJ = 25°C — TJ = 125°C
FIGURE 3 – CAPACITANCE
CAPACITANCE (pF)
REVERSE BIAS (VOLTS)
Cob
Cib
FIGURE 4 – CHARGE DATA
IC/IB = 10
VCC = 40 V
Q, CHARGE (pC)
QT
QA
IC, COLLECTOR CURRENT (mA)
FIGURE 5 – TURN-ON TIME
IC/IB = 10
TIME (ns)
tr @ VCC = 3.0 V
15 V
40 V
td @ VOB = 0
2.0 V
IC, COLLECTOR CURRENT (mA)
FIGURE 6 – RISE TIME
VCC = 40 V
IC/IB = 10
tr, RISE TIME (ns)
IC, COLLECTOR CURRENT (mA)
FIGURE 7 – STORAGE TIME
t's = ts – 1/8 tf
IB1 = IB2
IC/IB = 20
IC/IB = 10
IC/IB = 10
IC/IB = 20
t's, STORAGE TIME (ns)
IC, COLLECTOR CURRENT (mA)
FIGURE 8 – FALL TIME
IB1 = IB2
VCC = 40 V
IC/IB = 20
IC/IB = 20
IC/IB = 10
IC/IB = 10
tf, FALL TIME (ns)
IC, COLLECTOR CURRENT (mA)

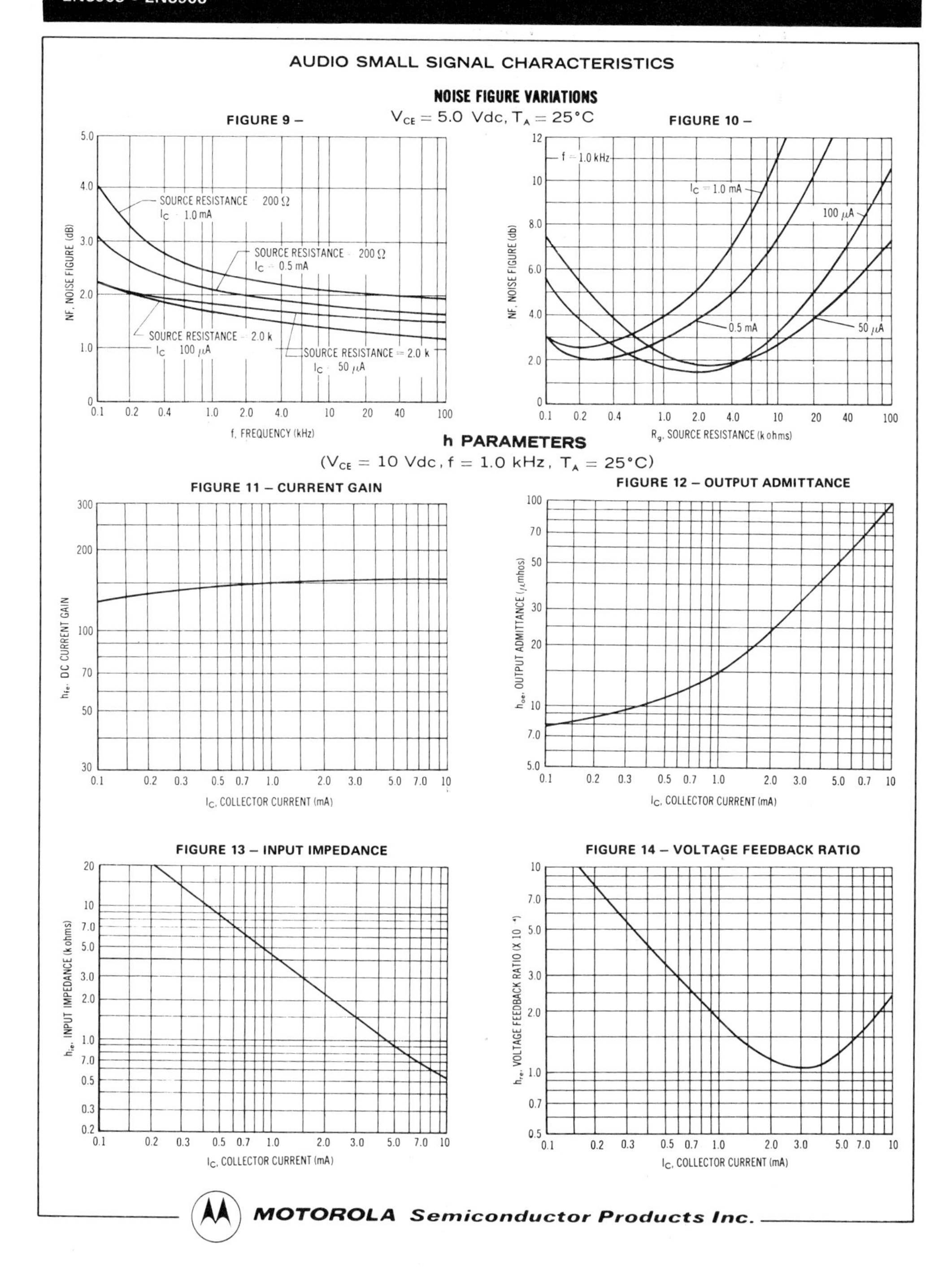
AUDIO SMALL SIGNAL CHARACTERISTICS
NOISE FIGURE VARIATIONS
V_{CE} = 5.0 Vdc, T_A = 25°C
FIGURE 9 –
SOURCE RESISTANCE = 200 Ω
I_C = 1.0 mA
SOURCE RESISTANCE = 200 Ω
I_C = 0.5 mA
SOURCE RESISTANCE = 2.0 k
I_C = 100 μA
SOURCE RESISTANCE = 2.0 k
I_C = 50 μA
NF, NOISE FIGURE (dB)
f, FREQUENCY (kHz)
FIGURE 10 –
f = 1.0 kHz
I_C = 1.0 mA
100 μA
0.5 mA
50 μA
NF, NOISE FIGURE (db)
R_g, SOURCE RESISTANCE (k ohms)
h PARAMETERS
(V_{CE} = 10 Vdc, f = 1.0 kHz, T_A = 25°C)
FIGURE 11 – CURRENT GAIN
h_{fe}, DC CURRENT GAIN
I_C, COLLECTOR CURRENT (mA)
FIGURE 12 – OUTPUT ADMITTANCE
h_{oe}, OUTPUT ADMITTANCE (μmhos)
I_C, COLLECTOR CURRENT (mA)
FIGURE 13 – INPUT IMPEDANCE
h_{ie}, INPUT IMPEDANCE (k ohms)
I_C, COLLECTOR CURRENT (mA)
FIGURE 14 – VOLTAGE FEEDBACK RATIO
h_{re}, VOLTAGE FEEDBACK RATIO (X 10⁻⁴)
I_C, COLLECTOR CURRENT (mA)
MOTOROLA Semiconductor Products Inc.

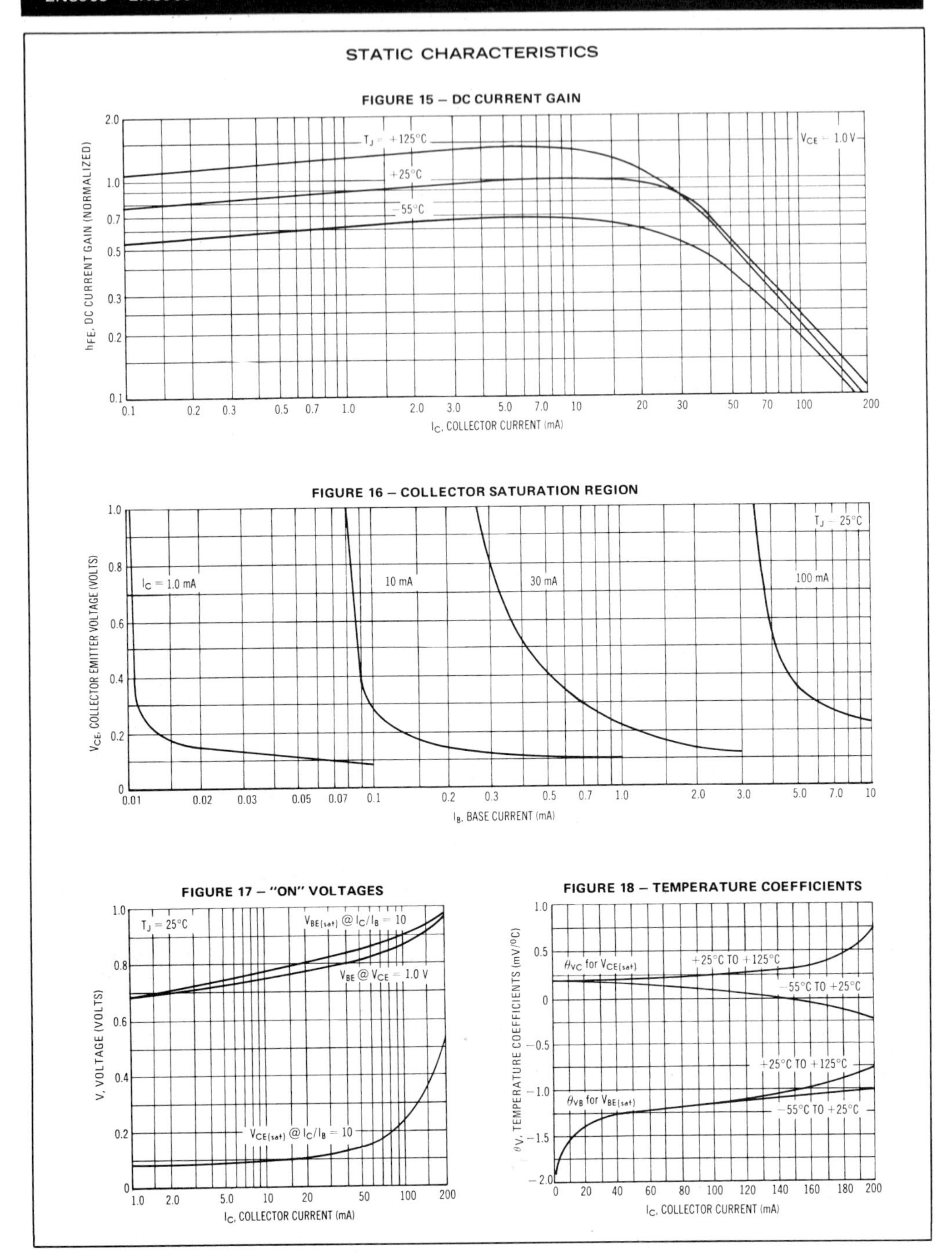
STATIC CHARACTERISTICS
FIGURE 15 – DC CURRENT GAIN
hFE, DC CURRENT GAIN (NORMALIZED)
TJ = +125°C
+25°C
−55°C
VCE = 1.0 V
2.0
1.0
0.7
0.5
0.3
0.2
0.1
0.1
0.2
0.3
0.5
0.7
1.0
2.0
3.0
5.0
7.0
10
20
30
50
70
100
200
IC, COLLECTOR CURRENT (mA)
FIGURE 16 – COLLECTOR SATURATION REGION
VCE, COLLECTOR EMITTER VOLTAGE (VOLTS)
TJ = 25°C
IC = 1.0 mA
10 mA
30 mA
100 mA
1.0
0.8
0.6
0.4
0.2
0
0.01
0.02
0.03
0.05
0.07
0.1
0.2
0.3
0.5
0.7
1.0
2.0
3.0
5.0
7.0
10
IB, BASE CURRENT (mA)
FIGURE 17 – "ON" VOLTAGES
V, VOLTAGE (VOLTS)
TJ = 25°C
VBE(sat) @ IC/IB = 10
VBE @ VCE = 1.0 V
VCE(sat) @ IC/IB = 10
1.0
0.8
0.6
0.4
0.2
0
1.0
2.0
5.0
10
20
50
100
200
IC, COLLECTOR CURRENT (mA)
FIGURE 18 – TEMPERATURE COEFFICIENTS
θV, TEMPERATURE COEFFICIENTS (mV/°C)
θVC for VCE(sat)
+25°C TO +125°C
−55°C TO +25°C
θVB for VBE(sat)
+25°C TO +125°C
−55°C TO +25°C
1.0
0.5
0
−0.5
−1.0
−1.5
−2.0
0
20
40
60
80
100
120
140
160
180
200
IC, COLLECTOR CURRENT (mA)

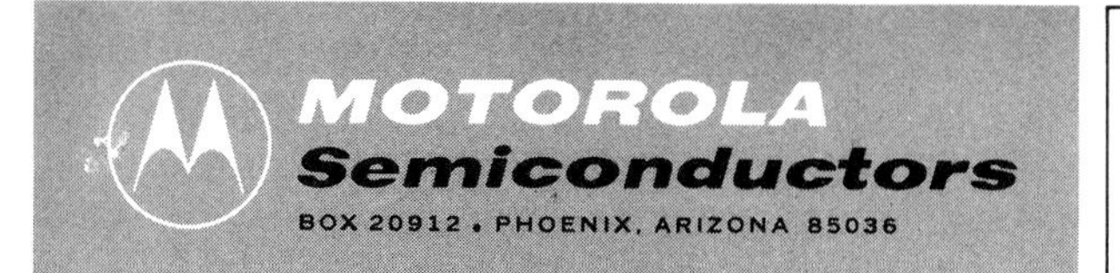

2N4441 thru 2N4444

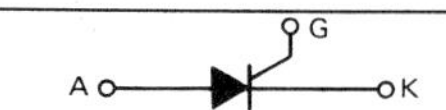

PLASTIC SILICON CONTROLLED RECTIFIERS

8.0 AMPERES RMS
50 thru 600 VOLTS

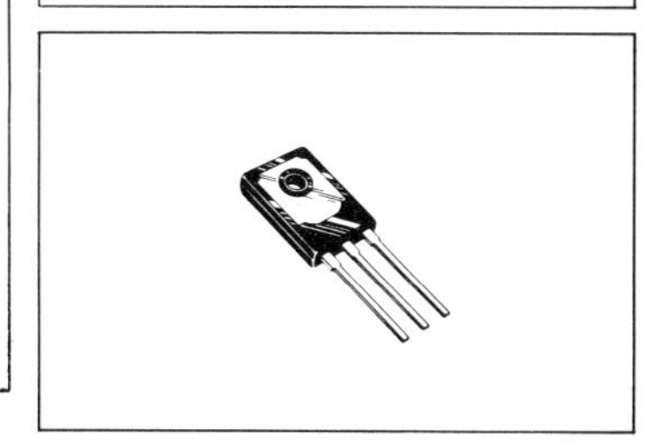

PLASTIC THYRISTORS

. . . designed for high-volume consumer phase-control applications such as motor speed, temperature, and light controls and for switching applications in ignition and starting systems, voltage regulators, vending machines, and lamp drivers requiring:

- Small, Rugged, Thermopad ▲ Construction — for Low Thermal Resistance, High Heat Dissipation, and Durability.
- Practical Level Triggering and Holding Characteristics @ 25°C
  I_{GT} = 7.0 mA (Typ)
  I_H = 6.0 mA (Typ)
- Low "On" Voltage — V_{TM} = 1.0 Volt (Typ) @ 5.0 Amp @ 25°C
- High Surge Current Rating — I_{TSM} = 80 Amp

MAXIMUM RATING (T_J = 100°C unless otherwise noted.)

Rating	Symbol	Value	Unit
*Repetitive Peak Reverse Blocking Voltage (Note 1)	V_{RRM}		Volts
2N4441		50	
2N4442		200	
2N4443		400	
2N4444		600	
*Non-Repetitive Peak Reverse Blocking Voltage (t = 5.0 ms (max) duration)	V_{RSM}		Volts
2N4441		75	
2N4442		300	
2N4443		500	
2N4444		700	
*RMS On-State Current (All Conduction Angles)	$I_{T(RMS)}$	8.0	Amp
Average On-State Current, T_C = 73°C	$I_{T(AV)}$	5.1	Amp
*Peak Non-Repetitive Surge Current (1/2 cycle, 60 Hz preceded and followed by rated current and voltage)	I_{TSM}	80	Amp
Circuit Fusing Considerations (T_J = −40 to +100°C; t = 1.0 to 8.3 ms)	I^2t	25	A^2s
*Peak Gate Power	P_{GM}	5.0	Watts
*Average Gate Power	$P_{G(AV)}$	0.5	Watt
*Peak Forward Gate Current	I_{GM}	2.0	Amp
*Peak Reverse Gate Voltage	V_{RGM}	10	Volts
*Operating Junction Temperature Range	T_J	-40 to +100	°C
*Storage Temperature Range	T_{stg}	-40 to +150	°C
Mounting Torque (6-32 screw) (Note 2)	—	8.0	in. lb.

THERMAL CHARACTERISTICS

Characteristic	Symbol	Typ	Max	Unit
*Thermal Resistance, Junction to Case	$R_{\theta JC}$	—	2.5	°C/W
Thermal Resistance, Junction to Ambient	$R_{\theta JA}$	40	—	°C/W

*Indicates JEDEC Registered Data.
▲Trademark of Motorola Inc.

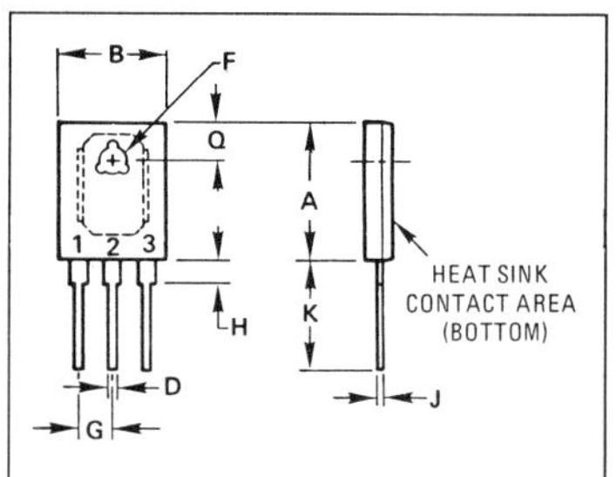

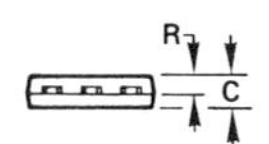

DIM	MILLIMETERS		INCHES	
	MIN	MAX	MIN	MAX
A	15.95	16.71	0.628	0.658
B	12.45	13.21	0.490	0.520
C	3.05	3.81	0.120	0.150
D	1.09	1.25	0.043	0.049
F	3.51	3.76	0.138	0.148
G	4.22 BSC		0.166 BSC	
H	—	3.18	—	0.125
J	0.76	0.86	0.030	0.034
K	14.99	16.51	0.590	0.650
Q	4.50	5.00	0.177	0.197
R	1.91	2.16	0.075	0.085

CASE 90-04

STYLE 1:
PIN 1. CATHODE
2. ANODE
3. GATE

 DS 6533 R1

ELECTRICAL CHARACTERISTICS (T_C = 25°C unless otherwise noted)

Characteristic	Symbol	Min	Typ	Max	Unit
*Peak Forward Blocking Voltage (T_J = 100°C) Note 1	V_{DRM}				Volts
2N4441		50	–	–	
2N4442		200	–	–	
2N4443		400	–	–	
2N4444		600	–	–	
Peak Forward Blocking Current (Rated V_{DRM}, T_J = 100°C, gate open)	I_{DRM}	–	–	2.0	mA
Peak Reverse Blocking Current (Rated V_{DRM}, T_J = 100°C, gate open)	I_{RRM}	–	–	2.0	mA
Gate Trigger Current (Continuous dc) (Anode Voltage = 7.0 Vdc, R_L = 100 Ohms)	I_{GT}				mA
T_C = 25°C		–	7.0	30	
* T_C = -40°C				60	
Gate Trigger Voltage (Continuous dc)	V_{GT}				Volts
(Anode Voltage = 7.0 Vdc, R_L = 100 Ohms) T_C = 25°C		–	0.75	1.5	
* (Anode Voltage = 7.0 Vdc, R_L = 100 Ohms) T_C = -40°C		–	–	2.5	
* (Anode Voltage = Rated V_{DRM}, R_L = 100 Ohms) T_J = 100°C		0.2	–	–	
Peak On-State Voltage (Pulse Width = 1.0 to 2.0 ms, Duty Cycle ≤ 2.0%)	V_{TM}				Volts
(I_{TM} = 5.0 A peak)		–	1.0	1.5	
* (I_{TM} = 15.7 A peak)		–	–	2.0	
Holding Current (Anode Voltage = 7.0 Vdc, gate open)	I_H				mA
T_C = 25°C		–	6.0	40	
* T_C = -40°C		–	–	70	
Gate Controlled Turn-On Time (I_{TM} = 5.0 A, I_{GT} = 20 mA)	t_{gt}	–	1.0	–	μs
Circuit Commutated Turn-Off Time	t_q				μs
(I_{TM} = 5.0 A, I_R = 5.0 A)		–	15	–	
(I_{TM} = 5.0 A, I_R = 5.0 A, T_J = 100°C)		–	20	–	
Critical Rate of Rise of Off-State Voltage (Rated V_{DRM}, Exponential Waveform, T_J = 100°C, Gate Open)	dv/dt	–	50	–	V/μs

*Indicates JEDEC Registered Data

Note 1. Ratings apply for zero or negative gate voltage but positive gate voltage shall not be applied concurrently with a negative potential on the anode. When checking forward or reverse blocking capability, thyristor devices should not be tested with a constant current source in a manner that the voltage applied exceeds the rated blocking voltage.

Note 2. Torque rating applies with use of torque washer (Shakeproof WD19522 #6 or equivalent). Mounting torque in excess of 8 in. lbs. does not appreciably lower case-to-sink thermal resistance. Anode lead and heatsink contact pad are common.

For soldering purposes (either terminal connection or device mounting), soldering temperatures shall not exceed +225°C.

MOTOROLA Semiconductor Products Inc. 

MOTOROLA
Semiconductors
BOX 955 • PHOENIX, ARIZONA 85001

THE RF LINE

MPF102

JUNCTION
FIELD-EFFECT
TRANSISTOR

SYMMETRICAL
SILICON
N-CHANNEL

SEPTEMBER 1966 — DS 5203

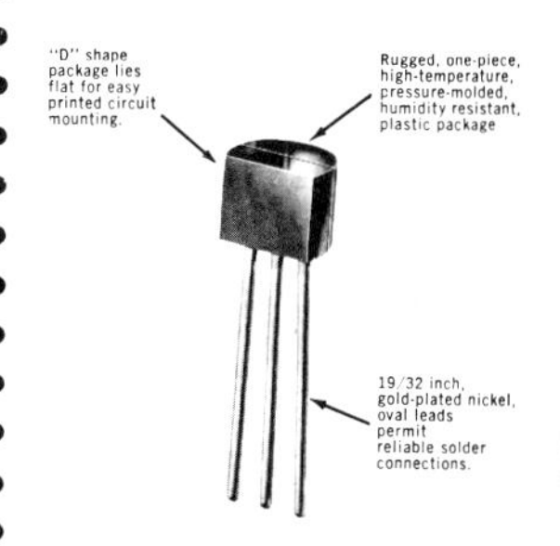

SILICON N-CHANNEL JUNCTION FIELD-EFFECT TRANSISTOR

. . . designed for VHF amplifier and mixer applications.

- Low Cross-Modulation and Intermodulation Distortion
- Guaranteed 100-MHz Parameters
- Drain and Source Interchangeable
- Low Transfer and Input Capacitance
- Low Leakage Current
- Unibloc* Plastic Encapsulated Package

MAXIMUM RATINGS ($T_A = 25°C$)

Characteristic	Symbol	Rating	Unit
Drain-Source Voltage	V_{DS}	25	Vdc
Drain-Gate Voltage	V_{DG}	25	Vdc
Gate-Source Voltage	V_{GS}	-25	Vdc
Gate Current	I_G	10	mAdc
Total Device Dissipation Derate above 25°C	P_D	200 2	mW mW/°C
Operating Junction Temperature	T_J	125	°C
Storage Temperature Range	T_{stg}	-65 to +150	°C

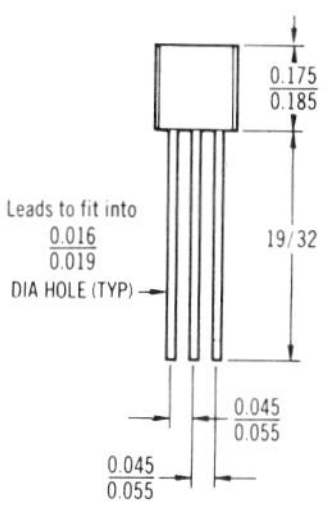

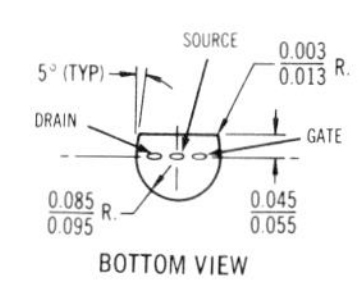

BOTTOM VIEW

TO-92

Drain and Source may be Interchanged.

*Trademark of Motorola Inc.

MOTOROLA Semiconductor Products Inc. A SUBSIDIARY OF MOTOROLA INC.

MPF102

ELECTRICAL CHARACTERISTICS (T_A = 25°C unless otherwise noted)

Characteristic	Symbol	Min	Max	Unit
OFF CHARACTERISTICS				
Gate-Source Breakdown Voltage (I_G = -10 μAdc, V_{DS} = 0)	BV_{GSS}	-25	—	Vdc
Gate Reverse Current (V_{GS} = -15 Vdc, V_{DS} = 0)	I_{GSS}	—	-2.0	nAdc
(V_{GS} = -15 Vdc, V_{DS} = 0, T_A = 100°C)		—	-2.0	μAdc
Gate-Source Cutoff Voltage (V_{DS} = 15 Vdc, I_D = 2.0 nAdc)	$V_{GS(off)}$	—	-8	Vdc
Gate-Source Voltage (V_{DS} = 15 Vdc, I_D = 0.2 mAdc)	V_{GS}	-0.5	-7.5	Vdc
ON CHARACTERISTICS				
Zero-Gate-Voltage Drain Current* (V_{DS} = 15 Vdc, V_{GS} = 0 Vdc)	I_{DSS}*	2	20	mAdc
DYNAMIC CHARACTERISTICS				
Forward Transfer Admittance* (V_{DS} = 15 Vdc, V_{GS} = 0, f = 1 kHz)	$\|y_{fs}\|$*	2000	7500	μmhos
Input Capacitance (V_{DS} = 15 Vdc, V_{GS} = 0, f = 1 MHz)	C_{iss}	—	7	pF
Reverse Transfer Capacitance (V_{DS} = 15 Vdc, V_{GS} = 0, f = 1 MHz)	C_{rss}	—	3	pF
Forward Transfer Admittance (V_{DS} = 15 Vdc, V_{GS} = 0, f = 100 MHz)	$\|y_{fs}\|$	1600	—	μmhos
Input Conductance (V_{DS} = 15 Vdc, V_{GS} = 0, f = 100 MHz)	$Re(y_{is})$	—	800	μmhos
Output Conductance (V_{DS} = 15 Vdc, V_{GS} = 0, f = 100 MHz)	$Re(y_{os})$	—	200	μmhos

*Pulse Test: Pulse Width ≤ 630 ms; Duty Cycle ≤ 10%

MOTOROLA Semiconductor Products Inc.

BOX 955 • PHOENIX, ARIZONA 85001 • A SUBSIDIARY OF MOTOROLA INC.

1815-1 PRINTED IN USA 3-67 IMPERIAL LITHO B1447

DS 5203

LM118/LM218/LM318 operational amplifier

general description

The LM118 series are precision high speed operational amplifiers designed for applications requiring wide bandwidth and high slew rate. They feature a factor of ten increase in speed over general purpose devices without sacrificing DC performance.

features

- 15 MHz small signal bandwidth
- Guaranteed 50V/μs slew rate
- Maximum bias current of 250 nA
- Operates from supplies of ±5V to ±20V
- Internal frequency compensation
- Input and output overload protected
- Pin compatible with general purpose op amps

The LM118 series has internal unity gain frequency compensation. This considerably simplifies its application since no external components are necessary for operation. However, unlike most internally compensated amplifiers, external frequency com pensation may be added for optimum performance For inverting applications, feedforward compen sation will boost the slew rate to over 150V/μs and almost double the bandwidth. Overcompensation can be used with the amplifier for greater stability when maximum bandwidth is not needed. Further, a single capacitor can be added to reduce the 0.1% settling time to under 1 μs.

The high speed and fast settling time of these op amps make them useful in A/D converters, oscillators, active filters, sample and hold circuits, or general purpose amplifiers. These devices are easy to apply and offer an order of magnitude better AC performance than industry standards such as the LM709.

The LM218 is identical to the LM118 except that the LM218 has its performance specified over a −25°C to +85°C temperature range The LM318 is specified from 0°C to +70°C.

schematic and connection diagrams

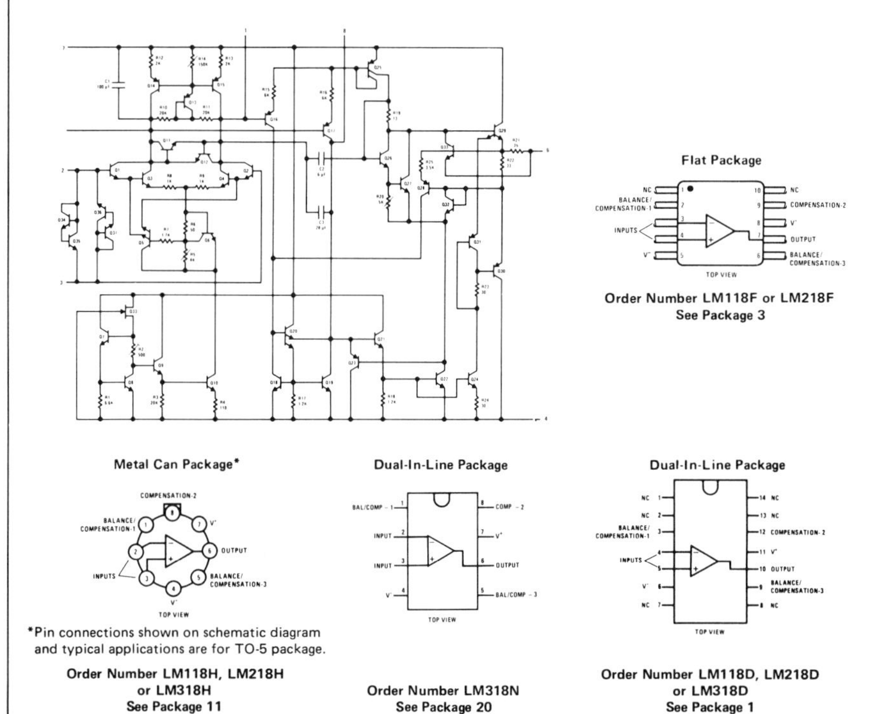

Order Number LM118F or LM218F
See Package 3

*Pin connections shown on schematic diagram and typical applications are for TO-5 package.

Order Number LM118H, LM218H or LM318H
See Package 11

Order Number LM318N
See Package 20

Order Number LM118D, LM218D or LM318D
See Package 1

absolute maximum ratings

Supply Voltage	±20V
Power Dissipation (Note 1)	500 mW
Differential Input Current (Note 2)	±10 mA
Input Voltage (Note 3)	±15V
Output Short-Circuit Duration	Indefinite
Operating Temperature Range	
LM118	−55°C to +125°C
LM218	−25°C to +85°C
LM318	0°C to +70°C
Storage Temperature Range	−65°C to +150°C
Lead Temperature (Soldering, 10 seconds)	300°C

electrical characteristics (Note 4)

PARAMETER	CONDITIONS	LM118/LM218			LM318			UNITS
		MIN	TYP	MAX	MIN	TYP	MAX	
Input Offset Voltage	$T_A = 25°C$		2	4		4	10	mV
Input Offset Current	$T_A = 25°C$		6	50		30	200	nA
Input Bias Current	$T_A = 25°C$		120	250		150	500	nA
Input Resistance	$T_A = 25°C$	1	3		0.5	3		MΩ
Supply Current	$T_A = 25°C$		5	8		5	10	mA
Large Signal Voltage Gain	$T_A = 25°C$, $V_S = \pm 15V$, $V_{OUT} = \pm 10V$, $R_L \geq 2\ k\Omega$	50	200		25	200		V/mV
Slew Rate	$T_A = 25°C$, $V_S = \pm 15V$, $A_V = 1$	50	70		50	70		V/μs
Small Signal Bandwidth	$T_A = 25°C$, $V_S = \pm 15V$		15			15		MHz
Input Offset Voltage				6			15	mV
Input Offset Current				100			300	nA
Input Bias Current				500			750	nA
Supply Current	$T_A = 125°C$		4.5	7				
Large Signal Voltage Gain	$V_S = \pm 15V$, $V_{OUT} = \pm 10V$, $R_L \geq 2\ k\Omega$	25			20			V/mV
Output Voltage Swing	$V_S = \pm 15V$, $R_L = 2\ k\Omega$	±12	±13		±12	±13		V
Input Voltage Range	$V_S = \pm 15V$	±11.5			±11.5			V
Common-Mode Rejection Ratio		80	100		70	100		dB
Supply Voltage Rejection Ratio		70	80		65	80		dB

Note 1: The maximum junction temperature of the LM118 is 150°C, the LM218 is 110°C, and the LM318 is 110°C. For operating at elevated temperatures, devices in the TO-5 package must be derated based on a thermal resistance of 150°C/W, junction to ambient, or 45°C/W, junction to case. For the flat package, the derating is based on a thermal resistance of 185°C/W when mounted on a 1/16-inch-thick epoxy glass board with ten, 0.03-inch-wide, 2-ounce copper conductors. The thermal resistance of the dual-in-line package is 100°C/W, junction to ambient.

Note 2: The inputs are shunted with back-to-back diodes for overvoltage protection. Therefore, excessive current will flow if a differential input voltage in excess of 1V is applied between the inputs unless some limiting resistance is used.

Note 3: For supply voltages less than ±15V, the absolute maximum input voltage is equal to the supply voltage.

Note 4: These specifications apply for $\pm 5V \leq V_S \leq \pm 20V$ and $-55°C \leq T_A \leq +125°C$, (LM118), $-25°C \leq T_A \leq +85°C$ (LM218), and $0°C \leq T_A \leq +70°C$ (LM318). Also, power supplies must be bypassed with 0.1μF disc capacitors.

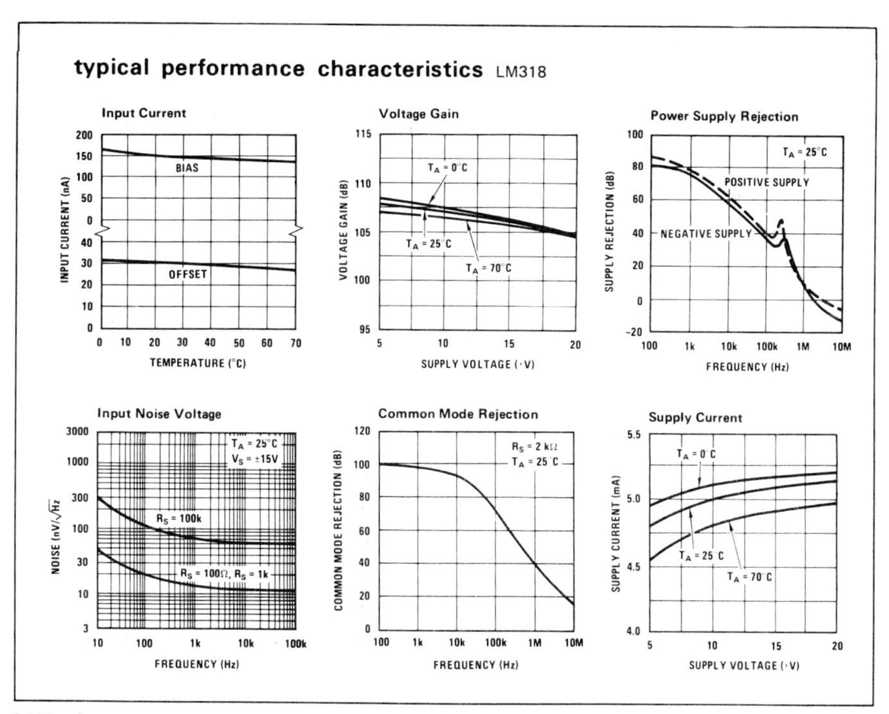

typical performance characteristics LM318
Input Current
INPUT CURRENT (nA)
BIAS
OFFSET
TEMPERATURE (°C)
Voltage Gain
VOLTAGE GAIN (dB)
T_A = 0°C
T_A = 25°C
T_A = 70°C
SUPPLY VOLTAGE (±V)
Power Supply Rejection
SUPPLY REJECTION (dB)
T_A = 25°C
POSITIVE SUPPLY
NEGATIVE SUPPLY
FREQUENCY (Hz)
Input Noise Voltage
NOISE (nV/√Hz)
T_A = 25°C
V_S = ±15V
R_S = 100k
R_S = 100Ω, R_S = 1k
FREQUENCY (Hz)
Common Mode Rejection
COMMON MODE REJECTION (dB)
R_S = 2 kΩ
T_A = 25°C
FREQUENCY (Hz)
Supply Current
SUPPLY CURRENT (mA)
T_A = 0°C
T_A = 25°C
T_A = 70°C
SUPPLY VOLTAGE (±V)

3-138

typical performance characteristics LM318 (Cont'd)

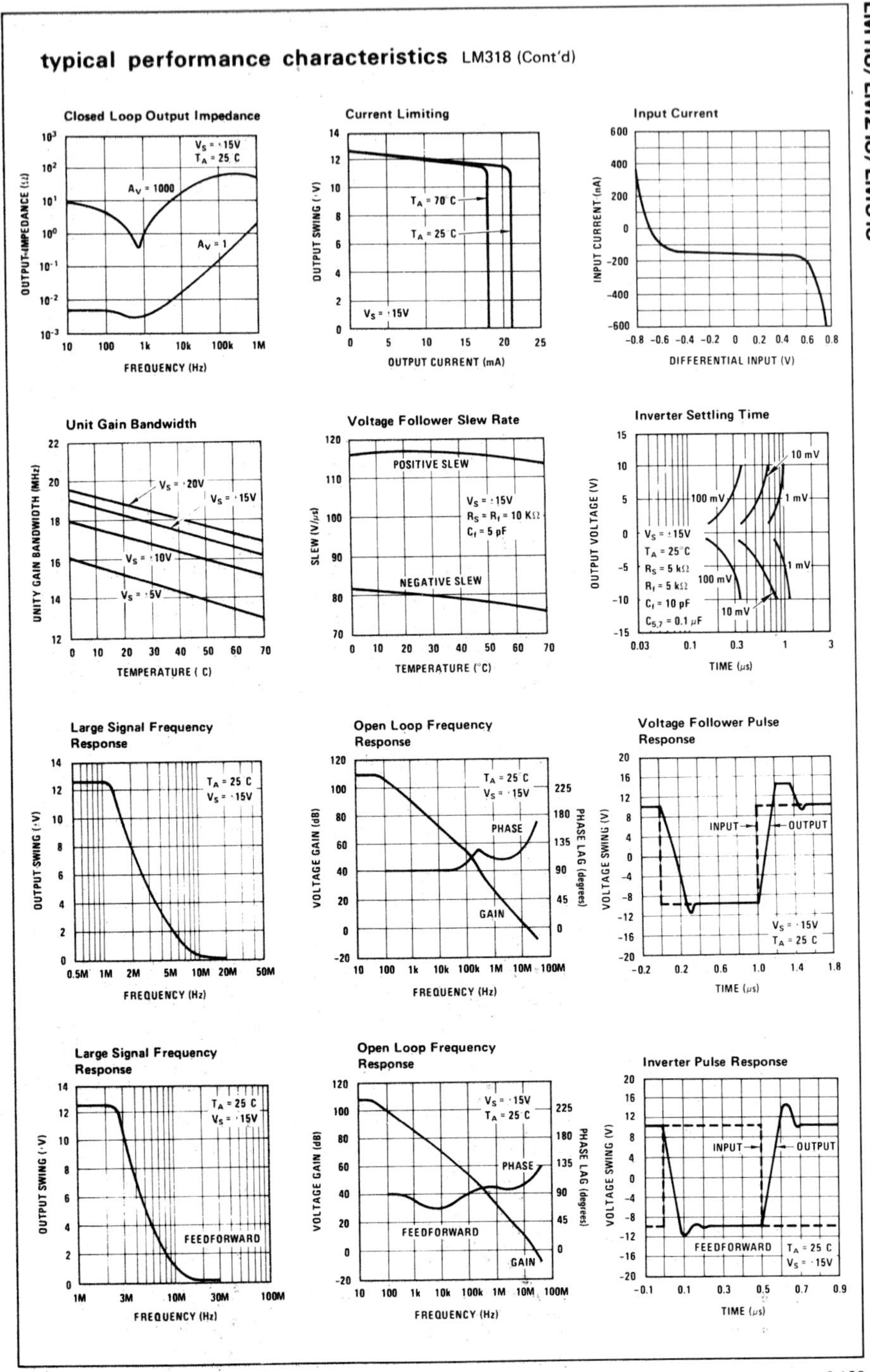

3

LM140A/LM140/LM340A/LM340 Series 3-Terminal Positive Regulators

General Description

The LM140A/LM140/LM340A/LM340 series of positive 3-terminal voltage regulators are designed to provide superior performance as compared to the previously available 78XX series regulator. Computer programs were used to optimize the electrical and thermal performance of the packaged IC which results in outstanding ripple rejection, superior line and load regulation in high power applications (over 15W).

With these advances in design, the LM340 is now guaranteed to have line and load regulation that is a factor of 2 better than previously available devices. Also, all parameters are guaranteed at 1A vs 0.5A output current. The LM140A/LM340A provide tighter output voltage tolerance, ±2% along with 0.01%/V line regulation and 0.3%/A load regulation.

Current limiting is included to limit peak output current to a safe value. Safe area protection for the output transistor is provided to limit internal power dissipation. If internal power dissipation becomes too high for the heat sinking provided, the thermal shutdown circuit takes over limiting die temperature.

Considerable effort was expended to make the LM140-XX series of regulators easy to use and minimize the number of external components. It is not necessary to bypass the output, although this does improve transient response. Input bypassing is needed only if the regulator is located far from the filter capacitor of the power supply.

Although designed primarily as fixed voltage regulators, these devices can be used with external components to obtain adjustable voltages and currents.

The entire LM140A/LM140/LM340A/LM340 series of regulators is available in the metal TO-3 power package and the LM340A/LM340 series is also available in the TO-220 plastic power package.

Features

- Complete specifications at 1A load
- Output voltage tolerances of ±2% at T_j = 25°C and ±4% over the temperature range (LM140A/LM340A)
- Fixed output voltages available 5, 6, 8, 10 12, 15, 18 and 24V
- Line regulation of 0.01% of V_{OUT}/V ΔV_{IN} at 1A load (LM140A/LM340A)
- Load regulation of 0.3% of V_{OUT}/A ΔI_{LOAD} (LM140A/LM340A)
- Internal thermal overload protection
- Internal short-circuit current limit
- Output transistor safe area protection

Typical Applications

Fixed Output Regulator

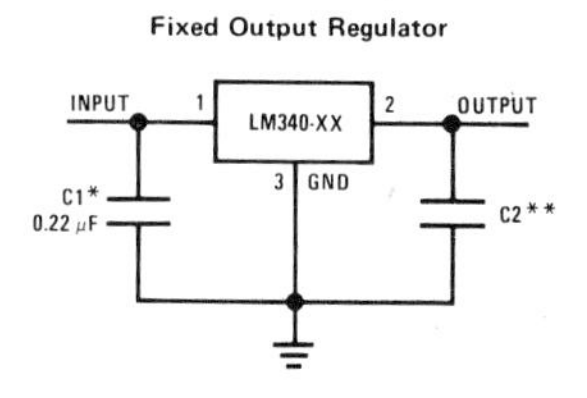

*Required if the regulator is located far from the power supply filter

**Although no output capacitor is needed for stability, it does help transient response. (If needed, use 0.1 µF, ceramic disc)

Adjustable Output Regulator

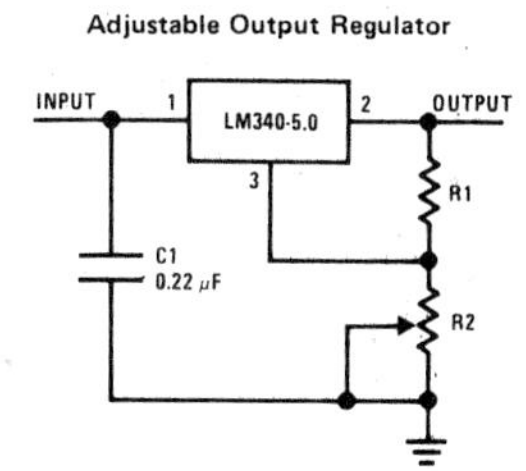

V_{OUT} = 5V + (5V/R1 + I_Q) R2
5V/R1 > 3 I_Q, load regulation (L_r) ≈ [(R1 + R2)/R1] (L_r of LM340-5)

Current Regulator

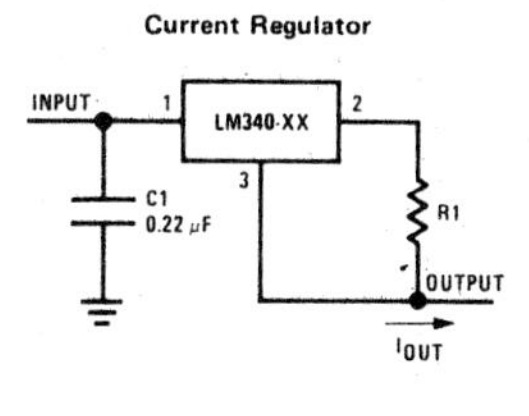

$$I_{OUT} = \frac{V2\text{-}3}{R1} + I_Q$$

ΔI_Q = 1.3 mA over line and load changes

Electrical Characteristics LM340 (Note 2)

0°C ≤ T_j ≤ +125°C unless otherwise noted.

OUTPUT VOLTAGE			5V			6V			8V			10V			12V			15V			18V			24V			
INPUT VOLTAGE (unless otherwise noted)			10V			11V			14V			17V			19V			23V			27V			33V			UNITS
	PARAMETER	CONDITIONS	MIN	TYP	MAX	MIN	TYP	MAX	MIN	TYP	MAX	MIN	TYP	MAX	MIN	TYP	MAX	MIN	TYP	MAX	MIN	TYP	MAX	MIN	TYP	MAX	
V_O	Output Voltage	T_j = 25°C, 5 mA ≤ I_O ≤ 1A	4.8	5	5.2	5.75	6	6.25	7.7	8	8.3	9.6	10	10.4	11.5	12	12.5	14.4	15	15.6	17.3	18	18.7	23.0	24	25.0	V
		P_D ≤ 15W, 5 mA ≤ I_O ≤ 1A	4.75		5.25	5.7		6.3	7.6		8.4	9.5		10.5	11.4		12.6	14.25		15.75	17.1		18.9	22.8		25.2	V
		V_{MIN} ≤ V_{IN} ≤ V_{MAX}		(7 ≤ V_{IN} ≤ 20)			(8 ≤ V_{IN} ≤ 21)			(10.5 ≤ V_{IN} ≤ 23)			(12.5 ≤ V_{IN} ≤ 25)			(14.5 ≤ V_{IN} ≤ 27)			(17.5 < V_{IN} ≤ 30)			(21 ≤ V_{IN} ≤ 33)			(27 ≤ V_{IN} ≤ 38)		V
ΔV_O	Line Regulation	I_O = 500 mA, T_j = 25°C		3	50		3	60		4	80		4	100		4	120		4	150		4	180		6	240	mV
		ΔV_{IN}		(7 ≤ V_{IN} ≤ 25)			(8 ≤ V_{IN} ≤ 25)			(10.5 ≤ V_{IN} ≤ 25)			(12.5 ≤ V_{IN} ≤ 25)			(14.5 ≤ V_{IN} ≤ 30)			(17.5 ≤ V_{IN} ≤ 30)			(21 ≤ V_{IN} ≤ 33)			(27 ≤ V_{IN} ≤ 38)		V
		I_O = 500 mA, 0°C ≤ T_j ≤ +125°C			50			60			80			100			120			150			180			240	mV
		ΔV_{IN}		(8 ≤ V_{IN} ≤ 20)			(9 ≤ V_{IN} ≤ 21)			(11 ≤ V_{IN} ≤ 23)			(13 ≤ V_{IN} ≤ 25)			(15 ≤ V_{IN} ≤ 27)			(18.5 ≤ V_{IN} ≤ 30)			(21.5 ≤ V_{IN} ≤ 33)			(28 ≤ V_{IN} ≤ 38)		V
		I_O ≤ 1A, T_j = 25°C			50			60			80			100			120			150			180			240	mV
		ΔV_{IN}		(7.3 ≤ V_{IN} ≤ 20)			(8.35 ≤ V_{IN} ≤ 21)			(10.5 ≤ V_{IN} ≤ 23)			(12.5 ≤ V_{IN} ≤ 25)			(14.6 ≤ V_{IN} ≤ 27)			(17.7 ≤ V_{IN} ≤ 30)			(21 < V_{IN} ≤ 33)			(27.1 ≤ V_{IN} ≤ 38)		V
		I_O ≤ 1A, 0°C ≤ T_j ≤ +125°C			25			30			40			50			60			75			90			120	mV
		ΔV_{IN}		(8 ≤ V_{IN} ≤ 12)			(9 ≤ V_{IN} ≤ 13)			(11 ≤ V_{IN} ≤ 17)			(14 ≤ V_{IN} ≤ 20)			(16 ≤ V_{IN} ≤ 22)			(20 ≤ V_{IN} ≤ 26)			(24 ≤ V_{IN} ≤ 30)			(30 ≤ V_{IN} ≤ 36)		V
ΔV_O	Load Regulation	T_j = 25°C, 5 mA ≤ I_O ≤ 1.5A		10	50		12	60		12	80		12	100		12	120		12	150		12	180		12	240	mV
		T_j = 25°C, 250 mA ≤ I_O ≤ 750 mA			25			30			40			50			60			75			90			120	mV
		5 mA ≤ I_O ≤ 1A, 0°C ≤ T_j ≤ +125°C			50			60			80			100			120			150			180			240	mV
I_Q	Quiescent Current	I_O ≤ 1A, T_j = 25°C			8			8			8			8			8			8			8			8	mA
		I_O ≤ 1A, 0°C ≤ T_j ≤ +125°C			8.5			8.5			8.5			8.5			8.5			8.5			8.5			8.5	mA
ΔI_Q	Quiescent Current Change	5 mA ≤ I_O ≤ 1A			0.5			0.5			0.5			0.5			0.5			0.5			0.5			0.5	mA
		T_j = 25°C, I_O ≤ 1A			1.0			1.0			1.0			1.0			1.0			1.0			1.0			1.0	mA
		V_{MIN} ≤ V_{IN} ≤ V_{MAX}		(7.5 ≤ V_{IN} ≤ 20)			(8.6 ≤ V_{IN} ≤ 21)			(10.6 ≤ V_{IN} ≤ 23)			(12.7 ≤ V_{IN} ≤ 25)			(14.8 ≤ V_{IN} ≤ 27)			(17.9 ≤ V_{IN} ≤ 30)			(21 ≤ V_{IN} ≤ 33)			(27.3 ≤ V_{IN} ≤ 38)		V
		I_O ≤ 500 mA, 0°C ≤ T_j ≤ +125°C			1.0			1.0			1.0			1.0			1.0			1.0			1.0			1.0	mA
		V_{MIN} ≤ V_{IN} ≤ V_{MAX}		(7 ≤ V_{IN} ≤ 25)			(8 ≤ V_{IN} ≤ 25)			(10.5 ≤ V_{IN} ≤ 25)			(12.5 ≤ V_{IN} ≤ 25)			(14.5 ≤ V_{IN} ≤ 30)			(17.5 ≤ V_{IN} ≤ 30)			(21 ≤ V_{IN} ≤ 33)			(27 ≤ V_{IN} ≤ 38)		V
V_N	Output Noise Voltage	T_A = 25°C, 10 Hz ≤ f ≤ 100 kHz		40			45			52			70			75			90			110			170		μV
$\frac{\Delta V_{IN}}{\Delta V_{OUT}}$	Ripple Rejection	f = 120 Hz, I_O ≤ 1A, T_j = 25°C or	62	80		59	78		56	76		55	74		55	72		54	70		53	69		50	66		dB
		f = 120 Hz, I_O ≤ 500 mA, 0°C ≤ T_j ≤ +125°C	62			59			56			55			55			54			53			50			dB
		V_{MIN} ≤ V_{IN} ≤ V_{MAX}		(8 ≤ V_{IN} ≤ 18)			(9 ≤ V_{IN} ≤ 19)			(11.5 ≤ V_{IN} ≤ 21.5)			(13.5 ≤ V_{IN} ≤ 23.5)			(15 ≤ V_{IN} ≤ 25)			(18.5 ≤ V_{IN} ≤ 28.5)			(22 ≤ V_{IN} ≤ 32)			(28 ≤ V_{IN} ≤ 38)		V
	Dropout Voltage	T_j = 25°C, I_{OUT} = 1A		2.0			2.0			2.0			2.0			2.0			2.0			2.0			2.0		V
R_O	Output Resistance	f = 1 kHz		8			9			12			16			18			19			22			28		mΩ
	Short-Circuit Current	T_j = 25°C		2.1			2.0			1.9			1.7			1.5			1.2			0.8			0.4		A
	Peak Output Current	T_j = 25°C		2.4			2.4			2.4			2.4			2.4			2.4			2.4			2.4		A
	Average TC of V_{OUT}	0°C ≤ T_j ≤ +150°C, I_O = 5 mA		-0.6			-0.7			-1.0			-1.2			-1.5			-1.8			-2.3			-3.0		mV/°C
V_{IN}	Input Voltage Required to Maintain Line Regulation	T_j = 25°C, I_O ≤ 1A	7.3			8.35			10.5			12.5			14.6			17.7			21			27.1			V

Note 2: All characteristics are measured with a capacitor across the input of 0.22 μF and a capacitor across the output of 0.1 μF. All characteristics except noise voltage and ripple rejection ratio are measured using pulse techniques (t_W ≤ 10 ms, duty cycle ≤ 5%). Output voltage changes due to changes in internal temperature must be taken into account separately.

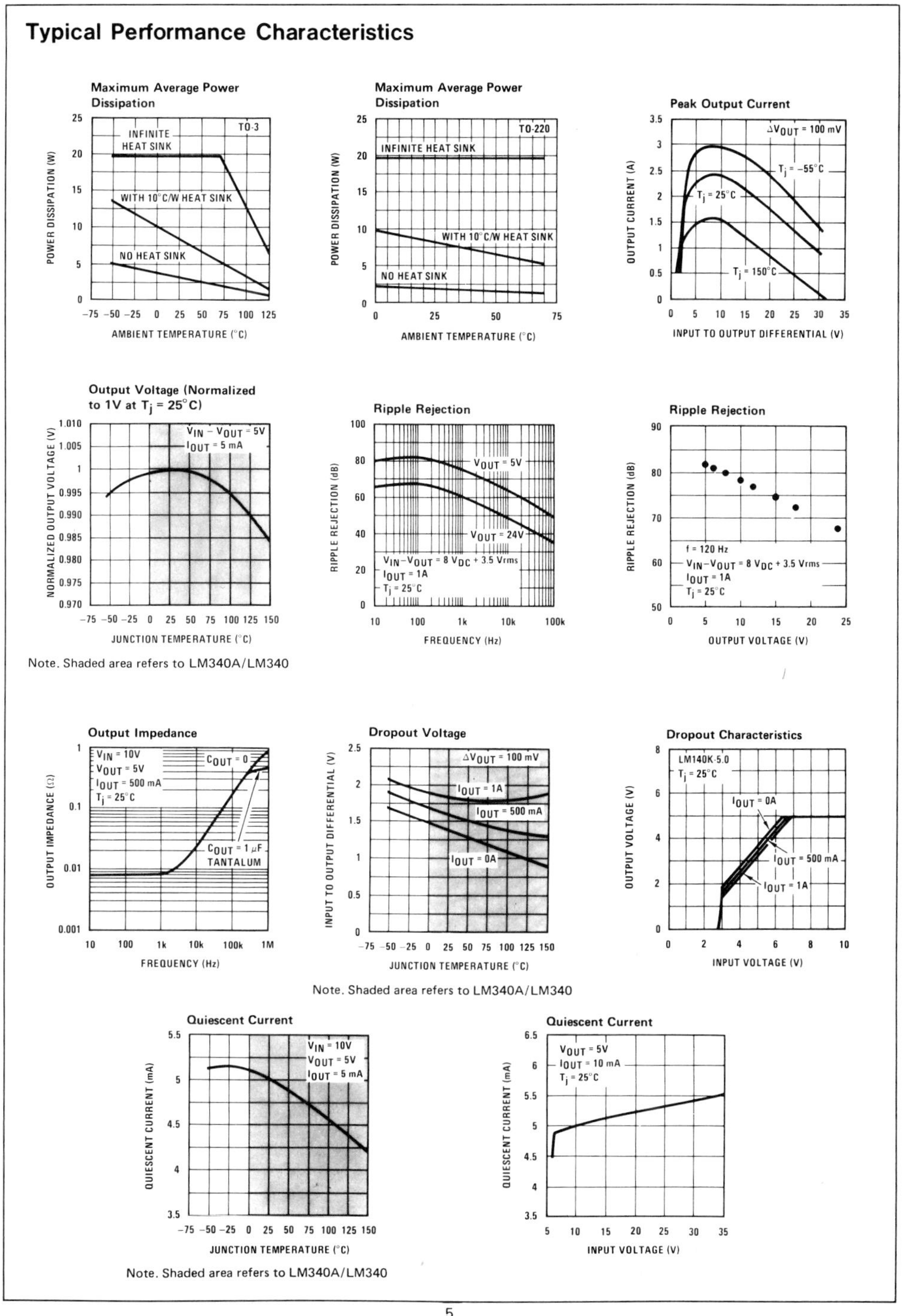
Typical Performance Characteristics
Maximum Average Power Dissipation
TO-3
INFINITE HEAT SINK
WITH 10°C/W HEAT SINK
NO HEAT SINK
POWER DISSIPATION (W)
AMBIENT TEMPERATURE (°C)
Maximum Average Power Dissipation
TO-220
INFINITE HEAT SINK
WITH 10°C/W HEAT SINK
NO HEAT SINK
POWER DISSIPATION (W)
AMBIENT TEMPERATURE (°C)
Peak Output Current
ΔVOUT = 100 mV
Tj = −55°C
Tj = 25°C
Tj = 150°C
OUTPUT CURRENT (A)
INPUT TO OUTPUT DIFFERENTIAL (V)
Output Voltage (Normalized to 1V at Tj = 25°C)
VIN − VOUT = 5V
IOUT = 5 mA
NORMALIZED OUTPUT VOLTAGE (V)
JUNCTION TEMPERATURE (°C)
Note. Shaded area refers to LM340A/LM340
Ripple Rejection
VOUT = 5V
VOUT = 24V
VIN−VOUT = 8 VDC + 3.5 Vrms
IOUT = 1A
Tj = 25°C
RIPPLE REJECTION (dB)
FREQUENCY (Hz)
Ripple Rejection
f = 120 Hz
VIN−VOUT = 8 VDC + 3.5 Vrms
IOUT = 1A
Tj = 25°C
RIPPLE REJECTION (dB)
OUTPUT VOLTAGE (V)
Output Impedance
VIN = 10V
VOUT = 5V
IOUT = 500 mA
Tj = 25°C
COUT = 0
COUT = 1 μF TANTALUM
OUTPUT IMPEDANCE (Ω)
FREQUENCY (Hz)
Dropout Voltage
ΔVOUT = 100 mV
IOUT = 1A
IOUT = 500 mA
IOUT = 0A
INPUT TO OUTPUT DIFFERENTIAL (V)
JUNCTION TEMPERATURE (°C)
Dropout Characteristics
LM140K-5.0
Tj = 25°C
IOUT = 0A
IOUT = 500 mA
IOUT = 1A
OUTPUT VOLTAGE (V)
INPUT VOLTAGE (V)
Note. Shaded area refers to LM340A/LM340
Quiescent Current
VIN = 10V
VOUT = 5V
IOUT = 5 mA
QUIESCENT CURRENT (mA)
JUNCTION TEMPERATURE (°C)
Note. Shaded area refers to LM340A/LM340
Quiescent Current
VOUT = 5V
IOUT = 10 mA
Tj = 25°C
QUIESCENT CURRENT (mA)
INPUT VOLTAGE (V)
5

Application Hints

The LM340 is designed with thermal protection, output short-circuit protection and output transistor safe area protection. However, as with *any* IC regulator, it becomes necessary to take precautions to assure that the regulator is not inadvertently damaged. The following describes possible misapplications and methods to prevent damage to the regulator.

Shorting the Regulator Input: When using large capacitors at the output of these regulators that have V_{OUT} greater than 6V, a protection diode connected input to output *(Figure 1)* may be required if the input is shorted to ground. Without the protection diode, an input short will cause the input to rapidly approach ground potential, while the output remains near the initial V_{OUT} because of the stored charge in the large output capacitor. The capacitor will then discharge through reverse biased emitter-base junction of the pass device, Q16, which breaks down at 6.5V and forward biases the base-collector junction. If the energy released by the capacitor into the emitter-base junction is large enough, the junction and the regulator will be destroyed. The fast diode in *Figure 1* will shunt the capacitor's discharge current around the regulator.

Raising the Output Voltage above the Input Voltage: Since the output of the LM340 does not sink current, forcing the output high can cause damage to internal low current paths in a manner similar to that just described in the "Shorting the Regulator Input" section.

Regulator Floating Ground *(Figure 2)*: When the ground pin alone becomes disconnected, the output approaches the unregulated input, causing possible damage to other circuits connected to V_{OUT}. If ground is reconnected with power "ON", damage may also occur to the regulator. This fault is most likely to occur when plugging in regulators or modules with on card regulators into powered up sockets. Power should be turned off first, or ground should be connected first if power must be left on.

Transient Voltages: If transients exceed the maximum rated input voltage of the 340, or reach more than 0.8V below ground and have sufficient energy, they will damage the regulator. The solution is to use a large input capacitor, a series input breakdown diode, a choke, a transient suppressor or a combination of these.

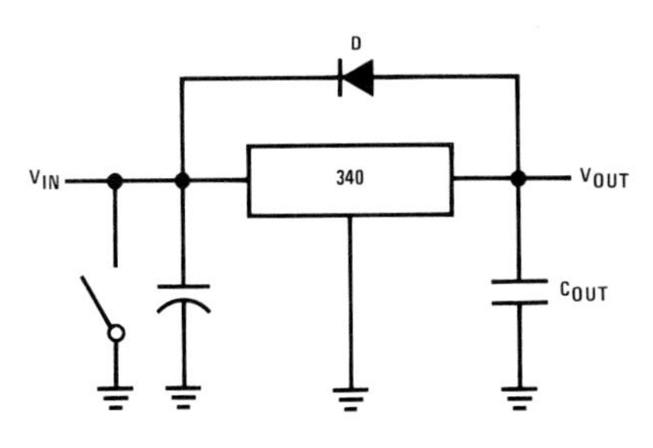

FIGURE 1. Input Short

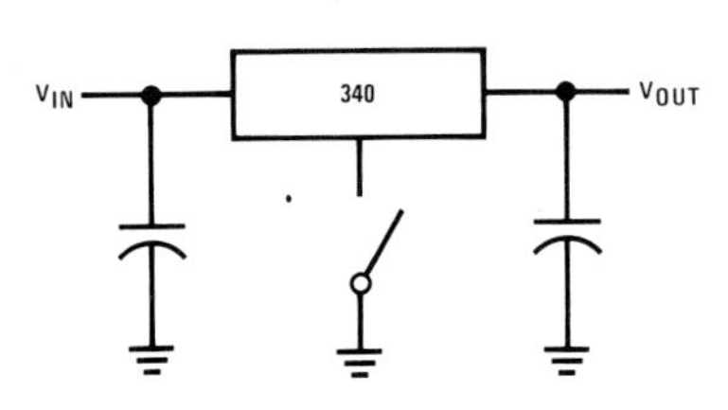

FIGURE 2. Regulator Floating Ground

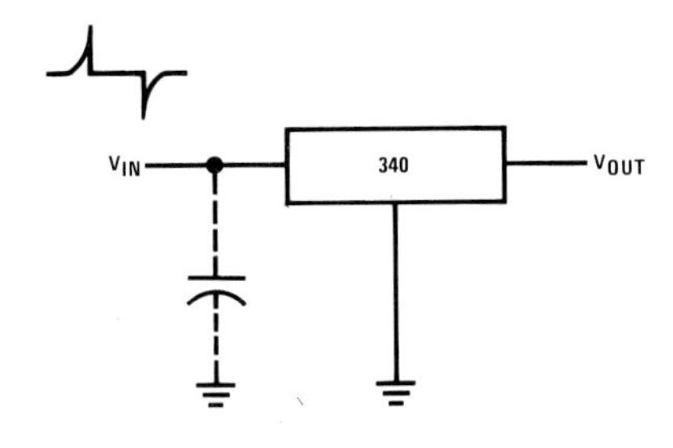

FIGURE 3. Transients

7

Connection Diagrams

Metal Can Package

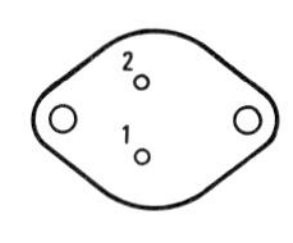

BOTTOM VIEW

Pin 1 — input
Pin 2 — output
Case — ground

Power Package

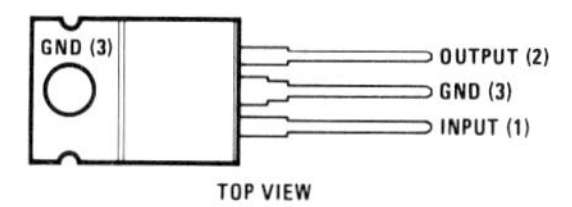

TOP VIEW

LM741/LM741A/LM741C/LM741E operational amplifier

general description

The LM741 series are general purpose operational amplifiers which feature improved performance over industry standards like the LM709. They are direct, plug-in replacements for the 709C, LM201, MC1439 and 748 in most applications.

The amplifiers offer many features which make their application nearly foolproof: overload protection on the input and output, no latch-up when the common mode range is exceeded, as well as freedom from oscillations.

The LM741C/LM741E are identical to the LM741/LM741A except that the LM741C/LM741E have their performance guaranteed over a 0°C to +70°C temperature range, instead of −55°C to +125°C.

schematic and connection diagrams (Top Views)

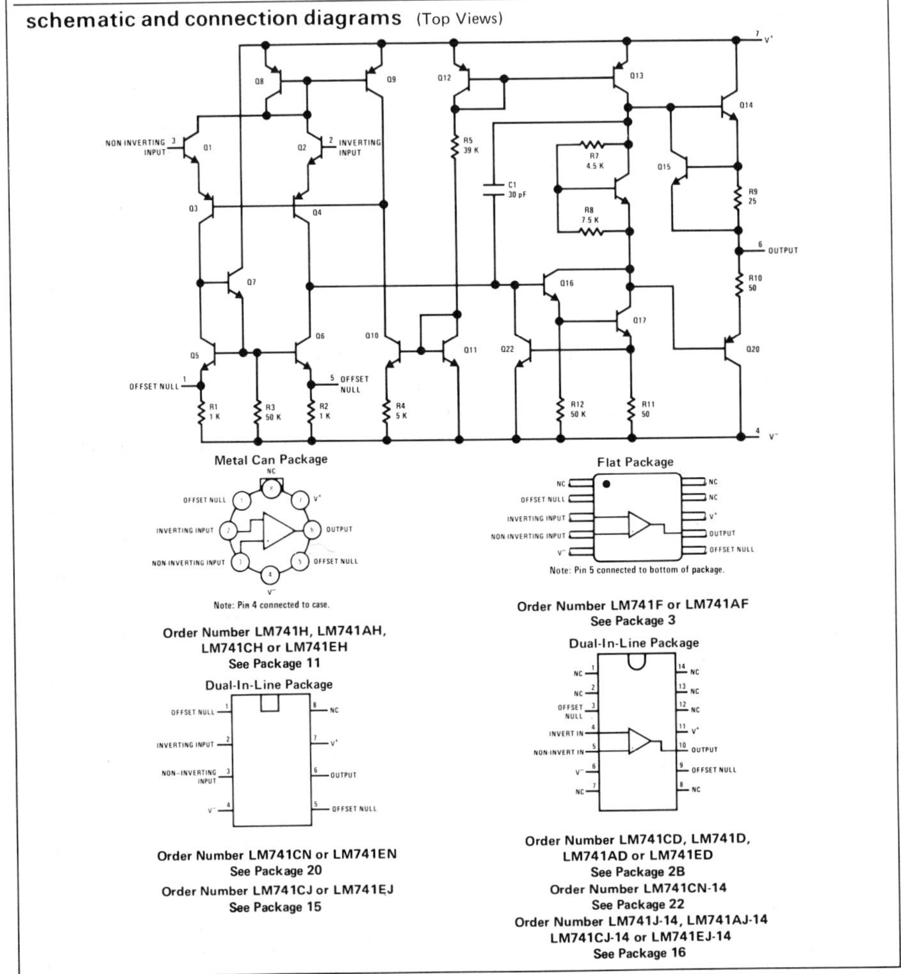

Order Number LM741H, LM741AH, LM741CH or LM741EH
See Package 11

Order Number LM741CN or LM741EN
See Package 20

Order Number LM741CJ or LM741EJ
See Package 15

Order Number LM741F or LM741AF
See Package 3

Order Number LM741CD, LM741D, LM741AD or LM741ED
See Package 2B

Order Number LM741CN-14
See Package 22

Order Number LM741J-14, LM741AJ-14 LM741CJ-14 or LM741EJ-14
See Package 16

absolute maximum ratings

	LM741A	LM741E	LM741	LM741C
Supply Voltage	±22V	±22V	±22V	±18V
Power Dissipation (Note 1)	500 mW	500 mW	500 mW	500 mW
Differential Input Voltage	±30V	±30V	±30V	±30V
Input Voltage (Note 2)	±15V	±15V	±15V	±15V
Output Short Circuit Duration	Indefinite	Indefinite	Indefinite	Indefinite
Operating Temperature Range	−55°C to +125°C	0°C to +70°C	−55°C to +125°C	0°C to +70°C
Storage Temperature Range	−65°C to +150°C	−65°C to +150°C	−65°C to +150°C	−65°C to +150°C
Lead Temperature (Soldering, 10 seconds)	300°C	300°C	300°C	300°C

electrical characteristics (Note 3)

PARAMETER	CONDITIONS	LM741A/LM741E			LM741			LM741C			UNITS
		MIN	TYP	MAX	MIN	TYP	MAX	MIN	TYP	MAX	
Input Offset Voltage	$T_A = 25°C$										
	$R_S \leq 10\ k\Omega$					1.0	5.0		2.0	6.0	mV
	$R_S \leq 50\Omega$		0.8	3.0							mV
	$T_{AMIN} \leq T_A \leq T_{AMAX}$										
	$R_S \leq 50\Omega$			4.0							mV
	$R_S \leq 10\ k\Omega$						6.0			7.5	mV
Average Input Offset Voltage Drift				15							μV/°C
Input Offset Voltage Adjustment Range	$T_A = 25°C$, $V_S = \pm 20V$	±10				±15			±15		mV
Input Offset Current	$T_A = 25°C$		3.0	30		20	200		20	200	nA
	$T_{AMIN} \leq T_A \leq T_{AMAX}$			70		85	500			300	nA
Average Input Offset Current Drift				0.5							nA/°C
Input Bias Current	$T_A = 25°C$		30	80		80	500		80	500	nA
	$T_{AMIN} \leq T_A \leq T_{AMAX}$			0.210			1.5			0.8	μA
Input Resistance	$T_A = 25°C$, $V_S = \pm 20V$	1.0	6.0		0.3	2.0		0.3	2.0		MΩ
	$T_{AMIN} \leq T_A \leq T_{AMAX}$, $V_S = \pm 20V$	0.5									MΩ
Input Voltage Range	$T_A = 25°C$							±12	±13		V
	$T_{AMIN} \leq T_A \leq T_{AMAX}$				±12	±13					V
Large Signal Voltage Gain	$T_A = 25°C$, $R_L \geq 2\ k\Omega$										
	$V_S = \pm 20V$, $V_O = \pm 15V$	50									V/mV
	$V_S = \pm 15V$, $V_O = \pm 10V$				50	200		20	200		V/mV
	$T_{AMIN} \leq T_A \leq T_{AMAX}$, $R_L \geq 2\ k\Omega$,										
	$V_S = \pm 20V$, $V_O = \pm 15V$	32									V/mV
	$V_S = \pm 15V$, $V_O = \pm 10V$				25			15			V/mV
	$V_S = \pm 5V$, $V_O = \pm 2V$	10									V/mV
Output Voltage Swing	$V_S = \pm 20V$										
	$R_L \geq 10\ k\Omega$	±16									V
	$R_L \geq 2\ k\Omega$	±15									V
	$V_S = \pm 15V$										
	$R_L \geq 10\ k\Omega$				±12	±14		±12	±14		V
	$R_L \geq 2\ k\Omega$				±10	±13		±10	±13		V
Output Short Circuit Current	$T_A = 25°C$	10	25	35		25			25		mA
	$T_{AMIN} < T_A \leq T_{AMAX}$	10		40							mA
Common-Mode Rejection Ratio	$T_{AMIN} \leq T_A \leq T_{AMAX}$										
	$R_S \leq 10\ k\Omega$, $V_{CM} = \pm 12V$				70	90		70	90		dB
	$R_S \leq 50\ k\Omega$, $V_{CM} = \pm 12V$	80	95								dB

electrical characteristics (con't)

PARAMETER	CONDITIONS	LM741A/LM741E			LM741			LM741C			UNITS
		MIN	TYP	MAX	MIN	TYP	MAX	MIN	TYP	MAX	
Supply Voltage Rejection Ratio	$T_{AMIN} \leq T_A \leq T_{AMAX}$, $V_S = \pm 20V$ to $V_S = \pm 5V$										
	$R_S \leq 50\Omega$	86	96								dB
	$R_S \leq 10\ k\Omega$				77	96		77	96		dB
Transient Response	$T_A = 25°C$, Unity Gain										
Rise Time			0.25	0.8		0.3			0.3		µs
Overshoot			6.0	20		5			5		%
Bandwidth (Note 4)	$T_A = 25°C$	0.437	1.5								MHz
Slew Rate	$T_A = 25°C$, Unity Gain	0.3	0.7			0.5			0.5		V/µs
Supply Current	$T_A = 25°C$					1.7	2.8		1.7	2.8	mA
Power Consumption	$T_A = 25°C$										
	$V_S = \pm 20V$		80	150							mW
	$V_S = \pm 15V$					50	85		50	85	mW
LM741A	$V_S = \pm 20V$										
	$T_A = T_{AMIN}$			165							mW
	$T_A = T_{AMAX}$			135							mW
LM741E	$V_S = \pm 20V$			150							mW
	$T_A = T_{AMIN}$			150							mW
	$T_A = T_{AMAX}$			150							mW
LM741	$V_S = \pm 15V$										
	$T_A = T_{AMIN}$					60	100				mW
	$T_A = T_{AMAX}$					45	75				mW

Note 1: The maximum junction temperature of the LM741/LM741A is 150°C, while that of the LM741C/LM741E is 100°C. For operation at elevated temperatures, devices in the TO-5 package must be derated based on a thermal resistance of 150°C/W junction to ambient, or 45°C/W junction to case. The thermal resistance of the dual-in-line package is 100°C/W junction to ambient. For the flat package, the derating is based on a thermal resistance of 185°C/W when mounted on a 1/16 inch thick epoxy glass board with ten, 0.03 inch wide, 2 ounce copper conductors.

Note 2: For supply voltages less than ±15V, the absolute maximum input voltage is equal to the supply voltage.

Note 3: Unless otherwise specified, these specifications apply for $V_S = \pm 15V$, $-55°C \leq T_A \leq +125°C$ (LM741/LM741A). For the LM741C/LM741E, these specifications are limited to $0°C \leq T_A \leq +70°C$.

Note 4: Calculated value from: BW (MHz) = 0.35/Rise Time(µs).